Digital Video FOR DUMMIES®

3RD EDITION

by Keith Underdahl

Wiley Publishing, Inc.

Digital Video For Dummies®, 3rd Edition

Published by
Wiley Publishing, Inc.
909 Third Avenue
New York, NY 10022
www.wiley.com

Copyright © 2003 by Wiley Publishing, Inc., Indianapolis, Indiana

Published by Wiley Publishing, Inc., Indianapolis, Indiana

Published simultaneously in Canada

No part of this publication may be reproduced, stored in a retrieval system or transmitted in any form or by any means, electronic, mechanical, photocopying, recording, scanning or otherwise, except as permitted under Sections 107 or 108 of the 1976 United States Copyright Act, without either the prior written permission of the Publisher, or authorization through payment of the appropriate per-copy fee to the Copyright Clearance Center, 222 Rosewood Drive, Danvers, MA 01923, (978) 750-8400, fax (978) 646-8700. Requests to the Publisher for permission should be addressed to the Legal Department, Wiley Publishing, Inc., 10475 Crosspoint Blvd., Indianapolis, IN 46256, (317) 572-3447, fax (317) 572-4447, e-mail: permcoordinator@wiley.com.

Trademarks: Wiley, the Wiley Publishing logo, For Dummies, the Dummies Man logo, A Reference for the Rest of Us!, The Dummies Way, Dummies Daily, The Fun and Easy Way, Dummies.com and related trade dress are trademarks or registered trademarks of Wiley Publishing, Inc., in the United States and other countries, and may not be used without written permission. All other trademarks are the property of their respective owners. Wiley Publishing, Inc., is not associated with any product or vendor mentioned in this book.

LIMIT OF LIABILITY/DISCLAIMER OF WARRANTY: WHILE THE PUBLISHER AND AUTHOR HAVE USED THEIR BEST EFFORTS IN PREPARING THIS BOOK, THEY MAKE NO REPRESENTATIONS OR WARRANTIES WITH RESPECT TO THE ACCURACY OR COMPLETENESS OF THE CONTENTS OF THIS BOOK AND SPECIFICALLY DISCLAIM ANY IMPLIED WARRANTIES OF MERCHANTABILITY OR FITNESS FOR A PARTICULAR PURPOSE. NO WARRANTY MAY BE CREATED OR EXTENDED BY SALES REPRESENTATIVES OR WRITTEN SALES MATERIALS. THE ADVICE AND STRATEGIES CONTAINED HEREIN MAY NOT BE SUITABLE FOR YOUR SITUATION. YOU SHOULD CONSULT WITH A PROFESSIONAL WHERE APPROPRIATE. NEITHER THE PUBLISHER NOR AUTHOR SHALL BE LIABLE FOR ANY LOSS OF PROFIT OR ANY OTHER COMMERCIAL DAMAGES, INCLUDING BUT NOT LIMITED TO SPECIAL, INCIDENTAL, CONSEQUENTIAL, OR OTHER DAMAGES.

For general information on our other products and services or to obtain technical support, please contact our Customer Care Department within the U.S. at 800-762-2974, outside the U.S. at 317-572-3993, or fax 317-572-4002.

Wiley also publishes its books in a variety of electronic formats. Some content that appears in print may not be available in electronic books.

Library of Congress Control Number: 2003105656

ISBN: 0-7645-4114-5

Manufactured in the United States of America

10 9 8 7 6 5 4 3 2

1B/RW/QX/QT/IN

is a trademark of Wiley Publishing, Inc.

About the Author

Keith Underdahl lives in Albany, Oregon. Professionally, Keith is an electronic-publishing specialist for Ages Software, where he serves as program manager, interface designer, video-media producer, multimedia specialist, graphic artist, and when the day is over, he even sweeps out the place. Mr. Underdahl has written numerous books, including *Teach Yourself Microsoft Word 2000*, *Microsoft Windows Movie Maker For Dummies*, *Macworld Final Cut Pro 2 Bible* (co-author), and *Adobe Premiere For Dummies*.

Dedication

Not all those who wander are lost.

— J. R. R. Tolkien

Author's Acknowledgments

When people find out that I have written several books, they usually react with awe and amazement, because writing an entire book seems like a monumental task. Writing a book *is* a monumental task, unless you're as fortunate and blessed as I am with such wonderful family and friends. I can't imagine getting up in the morning, much less writing a book like *Digital Video For Dummies*, without the help and support of my wife Christa. She has been my support staff, cheerleader, business manager, and inspiration throughout my writing career. Without Christa's initial encouragement and tireless ongoing support, I would not be writing.

Once again I must also thank my favorite movie subjects, Soren and Cole Underdahl. Not only do they take direction well, but they are also incredibly intelligent and look great on camera! Soren and Cole are featured in screen shots through this book, as well as some of the sample video clips found on the companion CD-ROM. Best of all, because they're my own boys, the Wiley Publishing legal department doesn't make me fill out all kinds of model releases and other daunting forms. If someone writes *Boilerplate For Dummies*, I'll buy the first copy.

Of course, I was working on this book before I even knew I was working on this book, making videos and developing my skills. I've received a lot of help from people in and out of the video and software business, including Jon D'Angelica, Patrick BeauLieu, John Bowne, Ingrid de la Fuente, havoc23, Linda Herd, Pete Langlois, Andy Marken, Rick Muldoon, Paulien Ruijssenaars, Steve from *Hard Times*, and probably a lot of other people I can't remember right at the moment.

Last but certainly not least, I wish to give my unwavering gratitude to the wonderful Composition Services team at Wiley Publishing. Basically I scribbled this whole book on a stack of cocktail napkins, and it was up to the folks at Wiley to turn it all into a coherent book. I wish to thank Steve Hayes, Linda Morris, Barry Childs-Helton, Dennis Short, and everyone else at Wiley. Keep up the great work, folks!

Publisher's Acknowledgments

We're proud of this book; please send us your comments through our online registration form located at www.dummies.com/register/.

Some of the people who helped bring this book to market include the following:

Acquisitions, Editorial, and Media Development

Project Editor: Linda Morris

Senior Acquisitions Editor: Steven Hayes

Senior Copy Editor: Barry Childs-Helton

Technical Editor: Dennis Short

Editorial Manager: Kevin Kirschner

Permissions Editor: Laura Moss

Media Development Specialist: Kit Malone

Media Development Supervisor: Richard Graves

Editorial Assistant: Amanda Foxworth

Cartoons: Rich Tennant www.the5thwave.com

Production

Project Coordinator: Erin Smith

Layout and Graphics: Seth Conley, Joyce Haughey, Leandra Hosier, Stephanie D. Jumper, Tiffany Muth, Janet Seib, Erin Zeltner

Proofreaders: John Tyler Connoley, Andy Hollandbeck, Carl William Pierce, Kathy Simpson, Brian H. Walls, **TECHBOOKS Production Services**

Indexer: TECHBOOKS Production Services

Publishing and Editorial for Technology Dummies

Richard Swadley, Vice President and Executive Group Publisher

Andy Cummings, Vice President and Publisher

Mary C. Corder, Editorial Director

Publishing for Consumer Dummies

Diane Graves Steele, Vice President and Publisher

Joyce Pepple, Acquisitions Director

Composition Services

Gerry Fahey, Vice President of Production Services

Debbie Stailey, Director of Composition Services

Contents at a Glance

Table of Contents

Introduction

If marketing folks want to breathe new life into an existing technology, the method *du jour* is to tack "digital" onto the name. Today we have *digital* cable, *digital* cell phones, *digital* fuel injection, and now, *digital* video. But unlike some other technologies that have recently earned the *digital* prefix, digital video isn't just a minor improvement over the old way of doing things. Digital video is a revolution that is changing the way we think about and use moving pictures.

Regular folks have had the capability to record their own video for many years now. Affordable film movie cameras have been available since the 1950s, and video cameras that record directly onto videotape have been with us for over two decades. But after you recorded some video or film with one of these old cameras, you couldn't do much else with it. You could show your movies to friends and family in raw, unedited form, but there was no confusing your rough home movie with a professional Hollywood production.

Digital camcorders provide a slight quality improvement over older camcorders, but the real advantage of digital video is that you can now easily edit your video on a computer. I don't have to tell you how far computer technology has progressed over the last few years, and you know that modern Macs and PCs can now do some pretty amazing things. In a matter of seconds, you can import video from your digital camcorder into your computer, cut out the scenes you don't want, add some special effects, and then instantly send your movies to friends over the Internet — or burn them to a DVD. The capability to easily edit your own movies adds a whole new level of creativity that was — just five years ago — the exclusive realm of broadcast and movie professionals.

In a culture so accustomed to and influenced by video images, it's actually kind of surprising that personal moviemaking hasn't burgeoned sooner. Video is the art of our time, and now — at last — you have the power to use this art for your own expression. What will you draw on your digital-video canvas?

Why This Book?

Digital video is a big, highly technical subject, so you need a guide to help you understand and use this technology. But you don't need a big book that is so highly technical that it just gathers dust on your bookshelf. You need easy-to-follow step-by-step instructions for the most important tasks, and you need tips and tricks to make your movies more successful. You need *Digital Video For Dummies, Third Edition.*

Needless to say, you're no "dummy." If you were, you wouldn't be reading this book and trying to figure out how to use digital video. Thanks to digital video, high-quality moviemaking has never been easier or more affordable. I have included instructions on performing the most important video-editing tasks, including lots of graphics so you can better visualize what I'm talking about. You'll also find tips and other ideas in this book that you wouldn't find in the documentation that comes with your editing software.

Digital Video For Dummies doesn't just help you use software or understand a new technology. It's about the art of moviemaking, and how you can apply this exciting new technology to make movies of your very own. I have designed this book to serve as a primer to moviemaking in general. Sections of this book will help you choose a good camcorder, shoot better video, publish movies online, and speak the industry technobabble like a Hollywood pro.

Foolish Assumptions

I've made a few basic assumptions about you as I have written this book. First, I assume that you have an intermediate knowledge of how to use a computer. Movie editing is one of the more technically advanced things you can do with a computer, so I assume that if you're ready to edit video, you already know how to locate and move files around on hard drives, open and close programs, and perform other such tasks. I assume that you are using either a Macintosh or a PC. In writing this book, I used both Mac and Windows software, and this book will be of use to you no matter which platform you use.

Another basic assumption I made is that you are not an experienced, professional moviemaker or video editor. I explain the fundamentals of videography and editing in ways that help you immediately get to work on your movie projects. Most of this book is based on the assumption that you are producing movies for fun or as a hobby. I also assume that you're not yet ready to spend many hundreds of dollars on highly advanced editing programs. In this book, I show you how to make amazing movies using software that is already installed on your computer (or that you can purchase for a modest sum). I have elected to show how to perform editing tasks primarily using Apple iMovie and Pinnacle Studio. iMovie is free for all Mac users, and Studio is a powerful yet affordable video-editing program for Windows. A trial version of Pinnacle Studio is included on the CD-ROM that accompanies this book.

Even if you are working in a professional environment and have just been tasked with creating your first company training or kiosk video, this book will help you grasp the fundamentals of digital video. Not only will this help you get to work quickly and efficiently, but I also include information to help you make an educated decision when your company gives you a budget to buy fancier editing software.

Conventions Used in This Book

Digital Video For Dummies helps you get started with moviemaking quickly and efficiently. Much of this book shows you how to perform tasks on your computer, which means you will find that this book is a bit different from other kinds of texts you have read. The following are some unusual conventions that you will encounter in this book:

- Filenames or lines of computer code will look like `THIS` or `this`. This style of print usually indicates something you should type in exactly as you see it in the book.
- Internet addresses will look something like this: `www.dummies.com`. Notice that we've left the `http://` part off the address because you never actually need to type it in your Web browser anymore.
- You will often be instructed to access commands from the menu bar of your video-editing program. The *menu bar* is that strip that lives along the top of the program window and usually includes menus like File, Edit, Window, and Help. If I'm telling you to access the Save command in the File menu, it will look like this: File ⇨ Save.
- You'll be using your mouse a lot. Sometimes you'll be told to click something to select it. This means that you should click *once* on whatever it is you are supposed to click, using the *left* mouse button if you use Microsoft Windows. Other times you will be told to *double-click* something; again, you double-click with the *left* mouse button if you are using Windows.

How This Book Is Organized

Believe it or not, I did put some forethought into the organization of this book. I hope you find it logically arranged and easy to use. This book is divided into six major parts. The parts are described in the following sections.

Part I: Getting Ready for Digital Video

You may be wondering: just what is this whole digital video thing, anyway? Part I introduces you to digital video. I'll show you what digital video is and what you can do with it. I'll also show you how to get your computer ready to work with digital video, and I'll help you choose a camcorder and other important moviemaking gear.

Part II: Gathering Footage

Editing video on your computer is just one part of the digital video experience. Before you can do any editing, you need something to actually edit. Part II shows you how to shoot better video, and then I show you how to get that video into your computer — even if you don't yet have a digital camcorder, or you just have some footage on old VHS tapes that you want to use. I also help you record and import better audio because good audio is just as important as video when you're making movies.

Part III: Editing Your Movie

Until just a few years ago, video editing was something that required professional-grade equipment, which cost in the hundreds of thousands of dollars. But with digital video and a semi-modern computer, editing video is now easy and very affordable. In Part III of this book, I'll introduce you to the basics of editing. You'll find out how to arrange scenes in the order you like and trim out the unwanted parts. I show you how to add cool transitions between video clips, use titles (text that appears onscreen), and top off your creation with sound effects, musical soundtracks, still graphics, and special effects.

Part IV: Sharing Your Video

When you've poured your heart into a movie project, you'll definitely want to share it with others. This part helps you share your movies on the Internet or on videotape. You even find out how to make your own DVDs in this part.

Part V: The Part of Tens

I wouldn't be able to call this a *For Dummies* book without a "Part of Tens" (really, it's in my contract). Actually, the Part of Tens always serves an important purpose. In this book, it gives me a chance to show you ten cool tips and tricks for better moviemaking, as well as ten tools that will improve your movies and make your work easier. Because there are a lot of video-editing programs out there to choose from, I also provide a chapter that compares ten of them, feature by feature.

Part VI: Appendixes

The appendixes provide quick, handy references on several important subjects. First I show you how to use the CD-ROM that accompanies this book.

Next up is a glossary to help you decrypt the alphabet soup of video-editing terms and acronyms. Additional appendixes help you install and update the editing programs shown throughout this book.

Icons Used in This Book

Occasionally you'll find some icons in the margins of this book. The text next to these icons includes information and tips that deserve special attention, and some warn you of potential hazards and pitfalls you may encounter. Icons you'll find in this book include

Although every word of *Digital Video For Dummies* is important (of course!), I sometimes feel the need to emphasize certain points or remind you of something that was mentioned elsewhere in the book. I use the Remember icon to provide this occasional emphasis.

Tips are usually brief instructions or ideas that, although not always documented, can greatly improve your movies and make your life easier. Tips are among the most valuable tidbits in this book.

Heed warnings carefully. Some warn of things that will merely inconvenience you, whereas others tell you when a wrong move could cause expensive and painful damage to your equipment and/or person.

Computer books are often stuffed with yards of technobabble, and if it's sprinkled everywhere, it can make the whole book a drag and just plain difficult to read. As much as possible I have tried to pull some of the deeply technical stuff out into these icons. This way, the information is easy to find if you need it, and just as easy to skip if you already have a headache.

The CD-ROM that accompanies this book contains sample clips and still images that you can use to practice the techniques we discuss, as well as a trial version of Pinnacle Studio. This icon lets you know when you might need to access something from the CD. For a full run-down on everything that appears on the CD, see Appendix A.

Where to Go From Here

You are about to enter the mad world of moviemaking. Exciting, isn't it? Digital video is *the* hot topic in technology today, and you're at the forefront of this multimedia revolution. If you're not sure whether your computer is ready for digital video, head on over to Chapter 2. If you still need to buy some gear or set up your movie studio, I suggest you visit Chapter 3. For tips on shooting better video with your new camcorder, spend some time in Chapter 4. Otherwise, you should jump right in and familiarize yourself with digital video, beginning with Chapter 1.

Part I
Getting Ready for Digital Video

"WELL! IT LOOKS LIKE SOMEONE FOUND THE 'LION'S ROAR' ON THE SOUND CONTROL PANEL."

In this part...

Films and videos have been around for many years now, but digital video is quickly changing how we think of movies. Never before has it been so easy for just about anyone to make high-quality movies with Hollywood-style features and effects. All you need is a camcorder and a semi-modern computer to make your own movie magic.

This first part of *Digital Video For Dummies, 3rd Edition* introduces you to digital video. You start out right away with the basics in Chapter 1, and I guide you through the process of making a simple movie. Subsequent chapters show you how to prepare your computer for working with digital video, and I review some of the other gear you'll need for moviemaking.

Chapter 1

Introducing Digital Video

In This Chapter

- Defining digital video
- Editing video on your own computer
- Sharing digital video

In 1996, I read a technical paper on a new technology from Apple Computer called *FireWire*. This new technology promised the ability to transfer data at speeds of up to 400 megabits per second. "Yeah, right!" I quietly scoffed to myself, "Why on Earth would anyone need to transfer that much data that quickly? Besides, Apple will be out of business by the end of '97."

Yeah, right.

Thankfully I was wrong about Apple, and I soon learned about a new phenomenon called *digital video* that could take advantage of this new FireWire technology. Digital video files are big, too big in fact for computers of just a few years ago to handle. But FireWire allows high quality video to be shared easily and efficiently between digital camcorders and computers.

Of course, more than just FireWire was needed for this digital video thing to catch on. Personal computers still had to become fast enough to handle digital video, and prices for digital camcorders only fell within reach of mere mortals just a couple of years ago. Digital video is here now, and anyone with a reasonably modern computer and a $500 digital camcorder can make movies like a pro. With the recent advent of DVD players and recordable DVD drives, sharing your high quality movies with others has never been easier.

This chapter introduces you to digital video and shows you how easy it is to edit and share your movies with others.

What Is Digital Video?

Human beings experience the world as an analog environment. When we take in the serene beauty of a rose garden, the mournful song of a cello, or the graceful motion of an eagle in flight, we are receiving a steady stream of infinitely variable data through our various senses. Of course, we don't think of all these things as "data" but rather as light, sound, smell, and touch.

Computers are pretty dumb compared to the human brain. They can't comprehend the analog data of the world; all computers understand are *yes* (one) and *no* (zero). In spite of this limitation, we force our computers to show pictures, play music, and display moving video; infinitely variable sounds, colors, and shapes must be converted into the language of computers — ones and zeros. This conversion process is called *digitizing*. Digital video — often abbreviated as *DV* — is video that has been digitized.

To fully understand the difference between analog data and digital data, suppose you want to draw the profile of a hill. An analog representation of the profile (shown in Figure 1-1) would follow the contour of the hill perfectly, because analog values are infinitely variable. However, a digital contour of that same hill would not be able to follow every single detail of the hill, because, as shown in Figure 1-2, digital values are made up of specifically defined, individual bits of data.

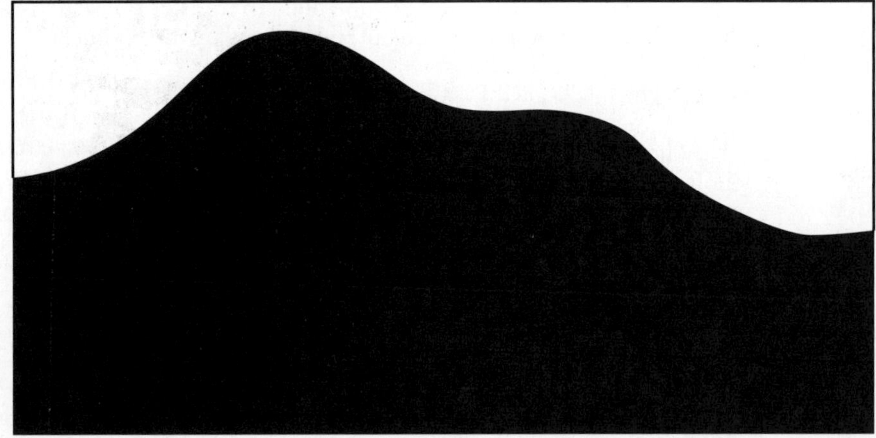

Figure 1-1: Analog data is infinitely variable.

Comparing analog and digital video

Digital recordings are theoretically inferior to analog recordings because analog recordings can contain more information. But the truth is that major advances in digital technology mean that this really doesn't matter. Yes, a digital recording must be made up of specific individual values, but modern recordings have so many discrete values packed so closely together that human eyes and ears can barely tell the difference. In fact, casual observation often reveals that digital recordings actually seem to be of a higher quality than analog recordings. Why?

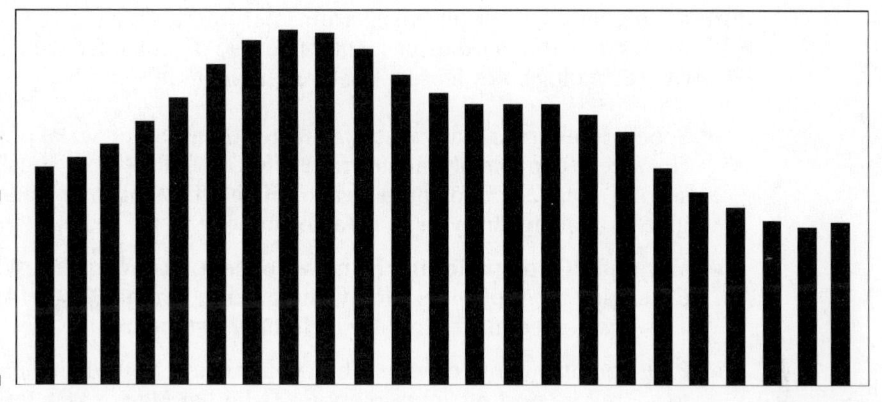

Figure 1-2: Digital data contains specific values.

A major problem with analog recordings is that they are highly susceptible to deterioration. Every time analog data is copied, some of the original, infinitely variable data is lost. This phenomenon, called *generational loss,* can be observed in that dark, grainy copy of a copy of a copy of a wedding video that was first shot more than 10 years ago. However, digital data doesn't have this problem. A one is always a one, no matter how many times it is copied, and a zero is always a zero. Likewise, analog recordings are more susceptible to deterioration after every playback, which explains why your 1964-vintage *Meet the Beatles* LP pops, hisses, and has lost many of its highs and lows over the years. Digital recordings are based on instructions that tell the computer how to create the data; as long as it can read the instructions, it creates the data the same way every time.

Whether you are editing analog or digital material, always work from a copy of the master and keep the master safe. When adding analog material to your project, the fewer generations your recording is from the original, the better.

When you consider the implications of generational loss on video editing, you begin to see what a blessing digital video really is. You're constantly copying, editing, and recopying content as you edit your movie projects — and with digital video, you can edit to your heart's content, confident that the quality won't diminish with each new copy you make.

Warming up to FireWire

FireWire is one of the hot new technologies that makes digital video so fun and easy to work with. FireWire — also sometimes called IEEE-1394 or i.LINK — was originally developed by Apple Computer and is actually an interface format for computer peripherals. Various peripherals including scanners, CD burners, external hard drives, and of course digital video cameras use FireWire technology. Key features of FireWire include

- **Speed:** FireWire is really fast, way faster than USB or serial ports. FireWire is capable of transfer rates up to 400Mbps (megabits per second). Digital video contains a lot of data that must be transferred quickly, making FireWire an ideal format.
- **Mac and PC compatibility:** (What a concept.) Although FireWire was developed by Apple, it is widely implemented in the PC world as well. This has helped make FireWire an industry standard.
- **Plug-and-play connectivity:** When you connect your digital camcorder to a FireWire port on your computer (whether Mac or PC), the camera is automatically detected. You won't have to spend hours installing software drivers or messing with obscure computer settings just to get everything working.
- **Device control:** Okay, this one isn't actually a feature *of* FireWire, it's just one of the things that makes using FireWire really neat. If your digital camcorder is connected to your computer's FireWire port, most video editing programs can control the camcorder's playback features. This means you don't have to juggle your fingers and try to press Play on the camcorder and Record in the software at exactly the same time. Just click Capture in a program like iMovie or Pinnacle Studio, and the software automatically starts and stops your camcorder as needed.
- **Hot-swap capability:** You can connect or disconnect FireWire components whenever you want. You don't need to shut down the computer, unplug power cables, or confer with your local public utility district before connecting or disconnecting a FireWire component.

All new Macintosh computers come with FireWire ports. Some — but not all — Windows PCs have FireWire ports as well. If your PC does not have a FireWire port, you can usually add one using an expansion card. (I cover all kinds of optional upgrades and adjustments to your system in Chapter 2.) Windows 98 and higher include software support for FireWire hardware. If you're buying a new PC and you plan to do a lot of video editing, consider a FireWire port a must-have feature.

All digital camcorders offer FireWire ports as well, although the port isn't always called FireWire. Sometimes FireWire ports are instead called "i.LINK" or simply "DV" by camcorder manufacturers who don't want to use Apple's trademarked FireWire name. But rest assured, all digital camcorders have a FireWire-compatible port. FireWire truly makes video editing easy, and if you are buying a new camcorder, I strongly recommend that you buy a camcorder that includes a FireWire port. Chapter 3 provides more detail on choosing a great digital camcorder.

Online versus offline editing

A video file represents a huge amount of information — so it takes up a lot of space in the digital world. You need fast hardware to handle video, and monster hard drives to store it. To conserve storage space during editing, professionals have long used a trick called *offline editing*. The idea is to capture lower-quality "working" copies of your video into the computer. After you complete all your edits and you're ready to make your final movie, the software decides which portions of the original video must be captured at full quality — and then automatically captures only the portions you need.

Conversely, if you work with full-quality video on your computer for all your edits, you are performing what video pros call *online editing*.

Offline and online editing are techie terms used by the pros. In practice, most affordable video editing programs don't give you many choices. A standout exception is Pinnacle Studio for Windows (as profiled in Chapter 5), which has an offline-editing feature called SmartCapture. This feature captures large sections of video at preview-quality — which means it may not look as sharp as full-quality video, but it doesn't take up nearly as much hard disk space. Then, when you're done editing, SmartCapture automatically captures only the full-quality footage needed for the movie and applies all your edits automatically.

Editing Video

Editing video projects with a program like Pinnacle Studio or Apple iMovie is pretty easy, but this wasn't always the case. Until recently, the only practical way for the average person to edit video was to connect two VCRs and use the Record and Pause buttons to cut out unwanted parts. This was a tedious and inefficient process. The up-to-date (and vastly improved) way to edit video is to use a computer — and the following sections introduce you to the video-editing techniques you're most likely to use.

Comparing editing methods

Video (and audio, for that matter) is considered a linear medium because it comes at you in a linear stream through time. A still picture, on the other hand, just sits there — you take it in all at once — and a Web site lets you jump randomly from page to page. Because neither of these is perceived as a stream, they're both examples of nonlinear media.

You tweak a linear medium (such as video) by using one of two basic methods — *linear* or *nonlinear* editing. If your approach to editing is linear, you must do all the editing work in chronological order, from the start of the movie to the finish. Here's an old-fashioned example: If you "edit" video by dubbing certain parts from a camcorder tape onto a VHS tape in your VCR, you have to do all your edits in one session, in chronological order. As you probably guessed, linear editing is terribly inefficient. If you dub a program and then decide to perform an additional edit, subsequent video usually has to be redubbed. (Oh, the pain, the tedium.)

What is the alternative? Thinking outside the line: *nonlinear editing*. You can do nonlinear edits in any order; you don't have to start at the beginning and slog on through to the end every time. The magical gizmo that makes nonlinear editing possible is a combination of the personal computer and programs designed for nonlinear editing (NLE). Using such a program (for example, Apple iMovie or Pinnacle Studio), you can navigate to any scene in the movie, insert scenes, move them around, cut them out of the timeline altogether, and slice, dice, tweak and fine-tune to your heart's content.

Editing a short video project

Editing video is really cool and easy to do if you have a reasonably modern computer. But why talk about editing when you can jump right into it? Here's the drill:

1. **Open Windows Movie Maker (Windows) or Apple iMovie (Macintosh).**

   If you don't know how to open your video-editing program, or if you aren't sure you have the latest version, check out Appendix C for information about iMovie, or Appendix E for the scoop on Windows Movie Maker. If you are prompted to create a new project by iMovie, create a new project and call it Chapter 1.

2. **Put the CD-ROM that accompanies this book in your CD-ROM drive.**
3. **Choose File⇨Import in iMovie or File⇨Import into Collections in Windows Movie Maker.**
4. **Browse to the `Samples\Chapter 1` folder on the CD-ROM.**

   In iMovie, hold down the ⌘ (Mac) key and click each clip once to select all three of them. In Windows Movie Maker, click the file `Chapter1` to select it.
5. **Click Open (iMovie) or Import (Windows Movie Maker).**

   Three clips appear in the browser window of your video-editing program, as shown in Figure 1-3. The figure shows iMovie, but Windows Movie Maker (shown in Figure 1-4) is fairly similar.
6. **Click-and-drag Clip 01 from the clip browser and then drop it on the storyboard.**
7. **Click-and-drag Clip 02 and drop it on the storyboard just after Clip 01.**

Figure 1-3: Three sample clips have been successfully imported.

Congratulations! You've just made your first movie edit. You should now have two clips on the storyboard, looking similar to Figure 1-4.

If your Windows Movie Maker window doesn't look quite like this, click the Show Timeline button (if you see it on-screen).

Well, okay, what's so nonlinear about that? After all, you placed one clip after another — that's about as linear as an edit can get. You could easily imagine doing the same thing with a camcorder, a VCR, and some cables.

Aha, but here's the kicker: What if you decide to insert Clip 03 in-between Clips 01 and 02? If you're "editing" with a camcorder and VCR, this move suddenly becomes a horrendously tricky edit to make. But with a nonlinear editing program like iMovie or Windows Movie Maker, the edit is easy. Just click-and-drag Clip 03 and drop it right between Clips 01 and 02. The software automatically shifts Clip 02 over to make room for Clip 03, as shown in Figure 1-5. Almost as easy as shuffling cards, edits like these are the essence of nonlinear video editing.

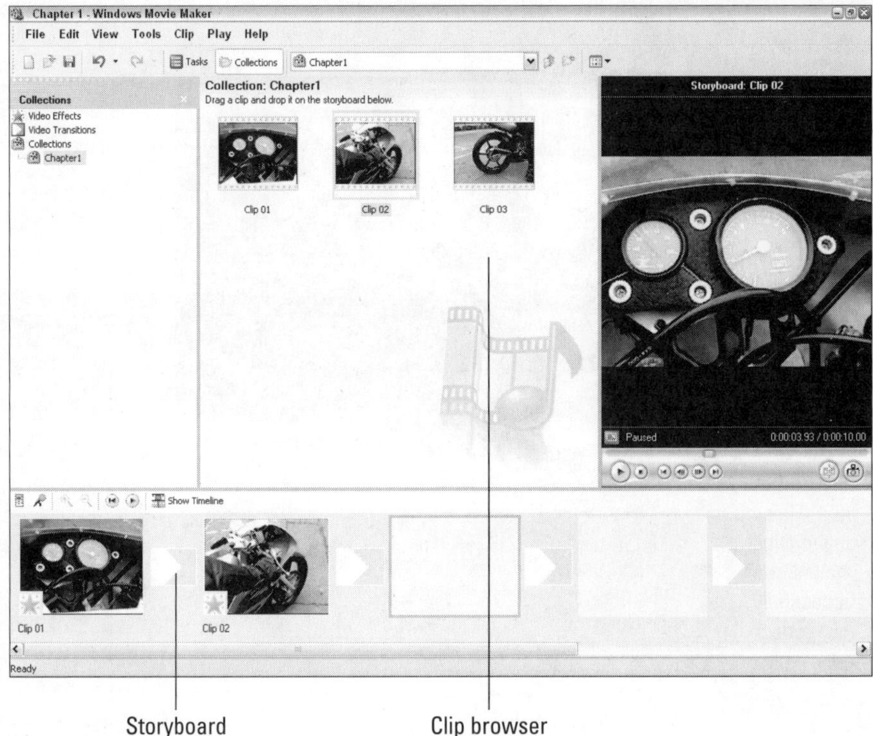

Figure 1-4: Two clips have been placed in the storyboard.

Performing Hollywood magic at the click of a mouse

The previous section shows the basics of making a movie by assembling clips in a specific order — and frankly, most of your editing work will probably consist of simple tasks like that. But when you want to go beyond ordinary, you can really spice up your movies by adding special effects or transitions between clips. (Special effects are covered in Chapter 11, and I show you just about everything you'll ever need to know about transitions in Chapter 9.)

Of course, there's no need to wait until later. Modern video-editing programs make it really easy to add special creative touches to your movies.

Creating a transition

You can add a transition to the simple movie you put together in the previous section by following these steps:

1. **Open the Chapter 1 sample project you created in the previous section if it isn't already open.**

   You can follow these steps using any movie project that includes two or more clips.

2. **Open the list of video transitions in your editing program.**

   In Apple iMovie, click the Trans button just below the browser window. In Windows Movie Maker, click Video Transitions under Collections on the left side of the screen.

3. **Click-and-drag one of the Circle transitions to a spot between two clips on the storyboard.**

   A transition indicator appears between the two clips, as shown in Figure 1-6.

4. **Click Play in the preview window to preview the transition.**

   If you are using iMovie, the transition may not appear immediately. If you see a tiny red progress bar under the transition, wait a few seconds for it to finish. When the progress bar is complete, you should be able to preview the transition.

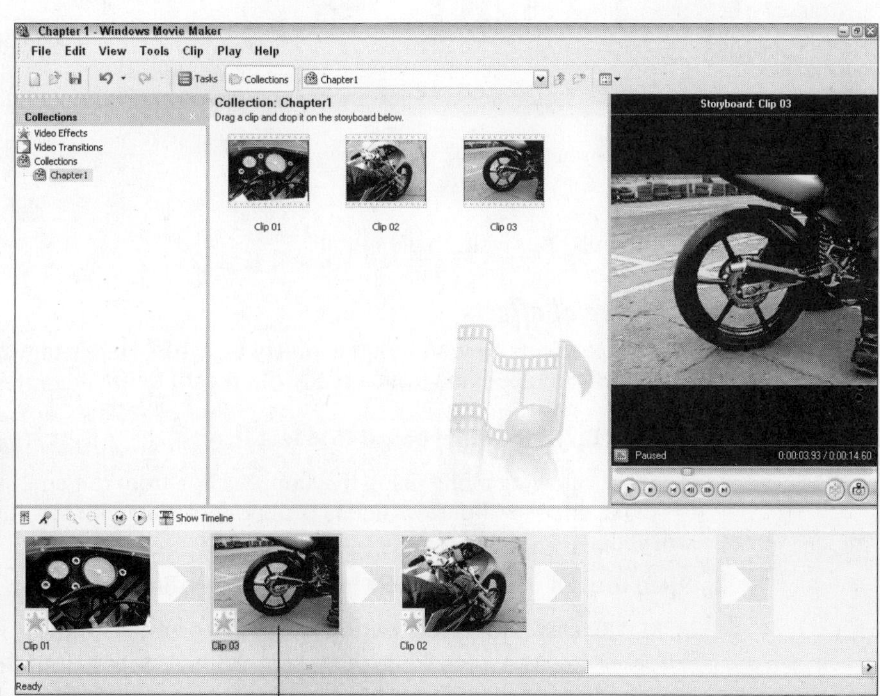

Figure 1-5: Clip 03 has been inserted between Clip 01 and Clip 02.

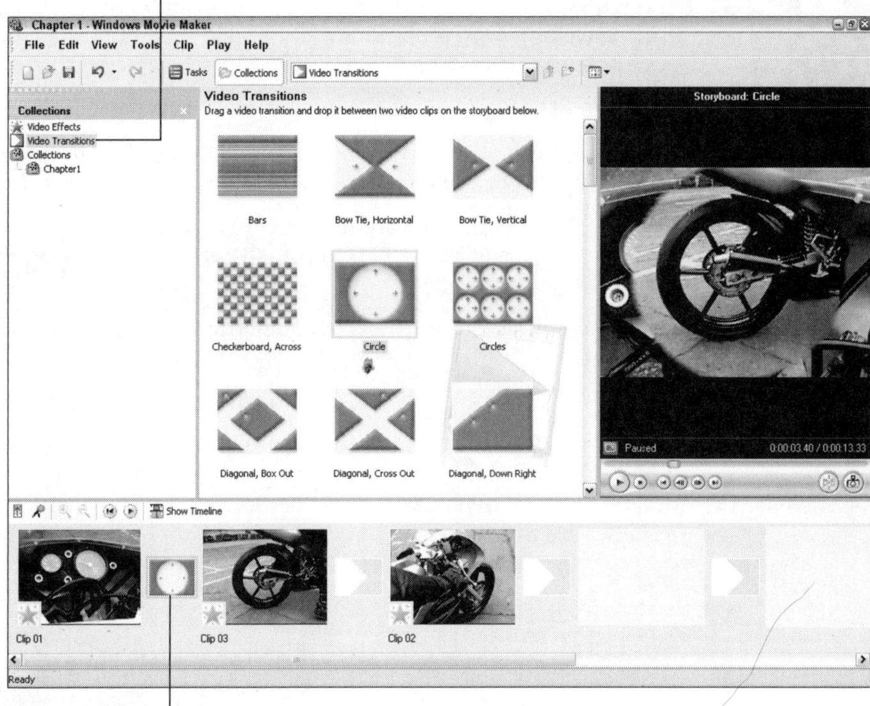

Figure 1-6: Transitions add a special touch to your movies.

Pretty cool, huh? But wait, that's not all!

Creating special effects

Adding special effects to your video is pretty easy too. Here's one that makes a video clip look like it came from a really old reel of film:

1. **Click a clip in the storyboard to select it.**

   If you're following along using the sample clips from the companion CD-ROM, choose Clip 02, which is probably the last clip in your storyboard.

2. **Open the list of video effects in your video-editing program.**

   In iMovie, click the Effects button under the browser window. In Windows Movie Maker, click Video Effects under Collections on the left side of the screen.

3. **Click an Aged Film effect to select it.**

   In iMovie, there is only one Aged Film effect. In Windows Movie Maker, scroll down in the list of effects and choose one of the Film Age effects. It doesn't matter if you choose Old, Older, or Oldest.

4. **Apply the effect to the clip.**

   In iMovie, click Apply at the top of the effects window. In Windows Movie Maker, click-and-drag the effect onto the clip on the storyboard.

5. **Click Play in the preview window to preview the effect as shown in Figure 1-7.**

   Again, if you're using iMovie, you will probably have to wait for the tiny red progress bar on the clip to finish before you can preview the effect.

These are just a couple of the cool things you can do with digital video. Part III of this book helps you explore the wonders of video editing in greater detail. So break out your director's chair and get ready to make some movie magic!

Figure 1-7: Making video look like aged film is just one of many special video effects you can use.

Sharing Video

One of the best things about digital video is that it enables you to get really creative with your own movie projects. To make your work worthwhile, you may want to share your video work with others. Thankfully, sharing digital video is pretty easy too. Part IV of this book shows you all the details of sharing video on tape, DVD, or the Internet, but the following sections provide a handy, brief glimpse of what you can do.

Exporting a movie

Modern video-editing programs are designed to make it as easy as possible to share your movie projects — often with no more than a couple of mouse clicks. For now, we'll export a movie that would be suitable for viewing over the Internet. This section uses the project created in Chapter 1 (using sample clips from the CD-ROM), but if you have your own edited movie, you can use it instead. The steps for exporting your movie are a little different depending on whether you are using Apple iMovie or Windows Movie Maker, so I'll address each program separately.

Exporting from Apple iMovie

Apple iMovie exports movies in QuickTime format, or you can export directly to your camcorder's videotape or Apple iDVD. To export your project in iMovie, follow these steps:

1. **Open the project you want to export (such as the** `Chapter 1` **project).**
2. **Choose File⇨Export.**

   The iMovie Export dialog box appears.
3. **Choose how you want to export your movie from the Export menu.**

   For now, I recommend that you choose To QuickTime (as shown in Figure 1-8).
4. **Choose a Format, such as Web.**
5. **Click Export.**
6. **Give a name for your movie file in the Save As box.**

   Make a note of the folder in which you are saving the movie. Choose a different folder if you wish.

   If you remove the `.mov` filename extension, Windows users (you probably know a few) will have a hard time viewing your movie.
7. **Click Save.**

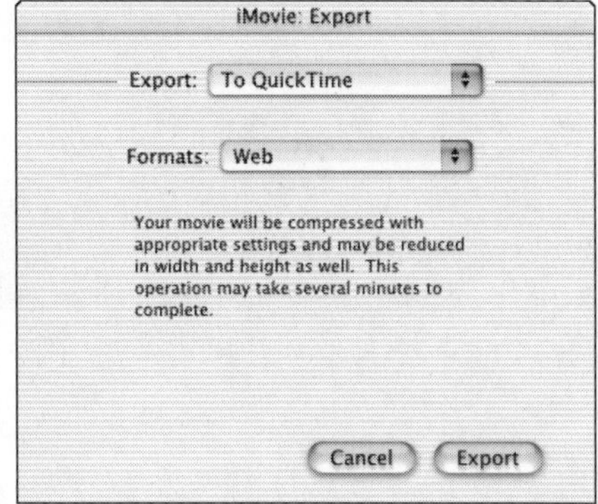

Figure 1-8: Choose export settings in this dialog box.

iMovie exports your movie. The export process may take a few minutes, depending on how long your movie is.

Exporting from Windows Movie Maker

Like iMovie, Windows Movie Maker also enables you to export video for a variety of applications. Windows Movie Maker is especially well suited to exporting movies for Internet playback. To export a movie for online viewing, follow these steps:

1. **Open the project that you want to export (such as the Chapter 1 project).**
2. **Choose File⇨Save Movie File.**

   The Save Movie Wizard appears.
3. **Choose an export format for your movie and then click Next.**

   For now I recommend choosing My Computer.
4. **Enter a filename for your movie and choose a location in which to save the file.**
5. **Click Next again.**

   The Save Movie Wizard shows details about the file, including the file size (see Figure 1-9).
6. **Click Next again.**

   The export process begins.
7. **When export is done, click Finish.**

Your movie will probably begin playing in Windows Media Player. Enjoy!

Click for more quality options.

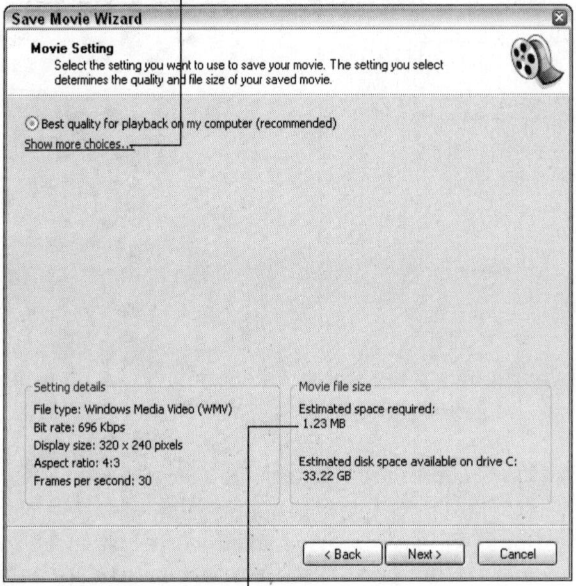

Estimated file size

Figure 1-9: Review movie details here.

Playing your movie

After your movie has been exported, playing it is pretty easy. Simply locate the file on your hard disk and double-click its name. The movie should automatically open and start to play, as shown in Figure 1-10.

If you exported your movie from Windows Movie Maker, the movie file will be in Windows Media (WMV) format. Despite the name, you don't have to be a Windows user to view Windows Media video. You do need Windows Media Player to view Windows Media files, but Microsoft offers a version of Windows Media Player for Macintosh. Figure 1-10 shows a Windows Media version of the Chapter 1 movie, playing contentedly on my Mac.

Movies created on a Mac are also cross-platform-friendly. iMovie outputs videos in Apple QuickTime format, and Windows versions of QuickTime (shown in Figure 1-11) have been available for years. Chapter 14 tells you more about available video-player programs.

Figure 1-10: Windows Media movies can be played on a Mac . . .

After you have previewed your movie, you can either share it with others or edit it some more. I usually go through the preview and re-edit process dozens of times before I decide that a movie project is ready for release, but thanks to digital video, re-editing is no problem at all. Chapter 13 offers some tips for previewing your movies more effectively.

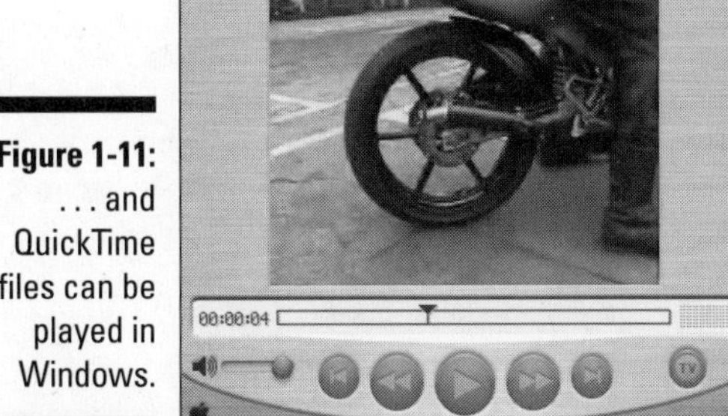

Figure 1-11: . . . and QuickTime files can be played in Windows.

Chapter 2

Getting Your Computer Ready for Digital Video

In This Chapter

- Deciding between a Mac or a PC
- Tips for safe computer upgrades
- Choosing and configuring a Mac for digital video
- Choosing and configuring a PC for digital video
- Selecting analog video capture hardware
- Improving the interface between you and your computer

A digital camcorder is just one part of the digital video equation. You also need a computer that is ready to capture, edit, and export digital video. Modern computers are pretty powerful, but there are still some important factors to consider if you want a computer well suited for digital video work. This chapter helps you choose a new computer, upgrade your current computer, and choose other gear that makes video editing fun and easy. But first, I'll start with something simple — trying to resolve the age-old debate of whether a Mac or Windows PC is better. (And if I succeed at that, I can work on a way to achieve world peace in the next chapter!)

Resolving the Mac-versus-PC Debate Once and for All

Yeah. Right.

Wanna start a ruckus? Wear an "I ♥ Bill Gates" t-shirt to a *Macworld* convention. Go ahead, I dare you.

Legal disclaimer: Wiley Publishing, Inc., is not responsible for physical or emotional harm which may result from compliance with the preceding foolish suggestion.

Like the debate over whether cats or dogs make better pets, the question of whether to use a Mac or a PC has been disputed tirelessly between the true believers. It has been a largely unproductive dispute: For the most part, Mac people are still Mac people, and PC people are still PC people.

But who is right? If you want the best computer for working with digital video, should you choose a PC or a Mac? Well, look at the important factors:

- **Ease of use:** Macintosh users often boast that their computers are exceedingly easy to use, and they are right. But if you're a long-time Windows user, you might not think so. Some things *are* easier to do on a Mac, but other things are easier to do in Windows. Neither system offers a clear advantage, so if you're a creature of habit, you'll probably be happiest if you stick with what you know.
- **Reliability:** The Windows Blue Screen of Death (you know, the dreaded screen that often appears when a Windows PC crashes) is world-famous and the butt of countless jokes. But the dirty little secret of the Macintosh world is that until recently, most Macs crashed nearly as often as Windows PCs. Apple's new Macintosh operating system — OS X — brings a new level of stability and refinement to the Macintosh world, but the latest Windows XP is pretty dependable as well. Reliability is important to you because video pushes your computer's performance to its limits. Get a Mac with OS X or a PC with Windows XP and you should be just fine.
- **Digital video support:** I can't be wishy-washy any longer; if you want a great computer ready to edit digital video right out of the box, a new Macintosh is the safer bet. All new Macs come with built-in FireWire ports, making it easy to hook up your digital camcorder. Macs also come with iMovie, a pretty good entry-level video-editing program. Windows comes with Windows Movie Maker, but it is not as capable as iMovie. Also, many Windows PCs still don't come with built-in FireWire, meaning you'll either have to special-order it or install a FireWire card yourself.

So there you have it: Macs and PCs are both pretty good. Sure, Macs all come with FireWire, but if you shop around, you should be able to find a Windows PC with FireWire for about the same price as a new Mac. Both platforms can make excellent video-editing machines, so if you're already dedicated to one or the other, you should be fine.

Just don't start any brawls, okay? You never know when a computer nerd wielding an iPod stylus might take offense.

Upgrading Your Computer

Picture this: Our hero carefully unscrews an access panel on the blinking device, revealing a rat's nest of wires and circuits. A drop of sweat runs down his face as the precious seconds tick away, and he knows that fate hangs by a slender thread. The hero's brow creases as he tries to remember the procedure: "Do I cut the blue wire or the red wire?"

If the thought of opening up your computer and performing upgrades fills you with a similar level of anxiety, you're not alone. The insides of modern computers can seem pretty mysterious, and you might be understandably nervous about tearing apart your expensive PC or Mac to perform hardware upgrades. Indeed, all the chips, circuit boards, and other electronic flotsam inside the computer case are sensitive and easily damaged. You can even hurt yourself if you're not careful, so if you don't have any experience with hardware upgrades, you are probably better off consulting a professional before making any changes or repairs to your PC.

But if you have done hardware upgrades before, digital video may well inspire you to make more upgrades now or in the near future. If you do decide to upgrade your computer, some basic rules include

- **Review your warranty.** Hardware upgrades might invalidate your computer's warranty if it still has one.
- **RTM.** This is geek-speak for, "Read The Manual." The owner's manual that came with your computer almost certainly contains important information about what can be upgraded and what can't. The manual may even have detailed, illustrated instructions for performing common upgrades.
- **Back up your data.** Back up your important files on recordable CDs, Zip disks, or another storage device available to you. You don't want to lose work, pictures, or other data that will be difficult or impossible to replace.
- **Gather license numbers and ISP (Internet Service Provider) information.** If you have any important things like software licenses stored in saved e-mails, print them out so that you have hard copies. Also, make sure that you have all the access information for your ISP (account name, password, dial-up numbers, server addresses, and so on) handy in case you need to re-install your Internet service.
- **Gather all your software CDs.** Locate all your original installation discs for your various programs, including your operating system (Mac OS or Windows), so that you'll be able to reinstall them later if necessary.
- **Turn off the power.** The computer's power should be turned off to avoid damage to components and electrocution to yourself.

- **Avoid static electricity build-up.** Even if you didn't just walk across a shag carpet and pet your cat, your body still probably has some static electricity built up inside. A tiny shock can instantly destroy the tiny circuits in expensive computer components. Before touching any components, touch your finger to a bare metal spot on your computer's case to ground yourself. I also recommend wearing a grounding strap, which can be purchased at most electronics stores for $2-3. Now *that's* what I call cheap insurance!
- **Handle with care.** Avoid touching chips and circuitry on the various computer components. Try to handle parts by touching only the edges or other less-delicate parts.
- **Protect those old parts.** If you are taking out an old component (such as a 64MB memory module) and replacing it with something better (like a 256MB memory module), the old part may still be worth something to somebody. If nothing else, if your newly purchased part is defective, at least you can put the old part back in to get your computer running again. And if the new part works fine, you may be able to salvage a few bucks by auctioning the old part on the Internet!

Again, when in doubt, you should consult with a computer hardware professional. In fact, you may find that the retailer that sold you the upgraded parts also offers low cost or even free installation service.

Using a Macintosh

Apple has put considerable effort into promoting the great multimedia capabilities of modern Macintosh computers. Indeed, Apple has been at the forefront of many important developments in digital video, and the video-capabilities of current Macs are impressive. In fact, any new Macintosh will work quite well with digital video. The following sections show you what to consider when choosing a new Mac, how to decide if your current Mac can handle digital video, and what video-editing software is available for your Mac.

Buying a new Mac

To work with digital video, your computer needs a powerful processor, lots of memory, a big hard drive, and a FireWire port. Any new Macintosh will meet these requirements. That said, not all new Macs are created equal. Generally speaking, the more money you spend, the better the Mac will be, so you shouldn't be surprised that some of the more affordable iMacs and iBooks are barely adequate. If you're serious about video, I recommend that your new Mac meet the following requirements:

- **512MB of RAM:** Video-editing software needs a lot of RAM (Random Access Memory) to work with, so the more the better. You might be able to get away with 256MB (megabytes) for a while, but you will find that video editing work is slow and tedious. Fortunately, the RAM in most Macs can be easily upgraded. In fact, iMacs incorporate a handy little access panel that enables you to upgrade RAM in mere seconds.
- **500MHz G3 processor:** A lot of video pros will tell you that you need a lot more, but you can get away with *only* a 500MHz (megahertz) G3 Mac. As you edit video, you'll find that you spend a lot of time sitting there, waiting for the computer to work. The faster the processor (and the more RAM), the less time you will spend waiting.
- **60GB hard drive:** I'm going to get e-mails from the video pros, I just know it! Video files take up lots and lots and lots of drive space. At one time, 60GB (gigabytes) was considered insanely massive for a hard drive, but when you are working with digital video, you'll eat up that space in a hurry. Less than five minutes of video eats up an entire gigabyte on your hard drive, which also has to hold software files, system files, and all kinds of other information. Lots of people will tell you that you need at least 100GB of drive space, and they're right: 100GB or more would be really nice. But you should be able to get away with a 60GB hard drive if your budget is tight. If you can afford a bigger drive, it's worth the expense.
- **OS X:** Any new Mac will include the latest version of Apple's operating system software. But if you are buying a previously enjoyed Mac, I strongly recommend that you buy one that already has OS version 10.1.5 or higher. If the computer you want to buy doesn't have OS X, factor the cost of a software upgrade (about $130) into the price.

If you're considering a portable iBook or PowerBook computer, pay special attention to the specifications before you buy. Portable computers usually have considerably less RAM and hard drive space than similarly priced iMacs and PowerMacs.

Upgrading ye olde Mac

If you already have a Mac that is a year or two old, you may still be able to use it with digital video if it meets the basic requirements described in the previous section. If it doesn't meet those requirements, you might be able to upgrade it. As a general rule, however, if your Mac doesn't already have a G3 or higher processor, upgrading is probably going to be more expensive or challenging than simply buying a new Mac. And in the end, a very old upgraded Mac probably will not perform that well anyway.

One of the biggest obstacles you'll face involves FireWire. If your Mac does not already have a FireWire port, you may have difficulty adding one. PowerMacs can usually be upgraded with a FireWire card, but the few Mac-compatible FireWire cards available tend to be pretty expensive. The Media 100 EditDV 2.0 FireWire card, for example, will set you back about $580. If you have a PowerBook G3, a slightly more affordable option is the Digital Origin MotoDV Mobile (retail price about $300) editing suite with a FireWire card that uses the PowerBook's CardBus interface.

Before you think you can get away without FireWire, keep in mind that if your Mac is too old for FireWire, its USB port won't be fast enough to capture full-quality digital video either.

Some parts of your Mac may be more easily upgradeable, as described in the following sections. For more details on upgrade information and installation instructions, check out *The iMac For Dummies* by David Pogue, or *Mac OS X All-in-One Desk Reference For Dummies* by Mark Chambers (both published by Wiley Publishing, Inc.). Macworld (`www.macworld.com`) also provides online articles and tutorials to help you upgrade your Mac. And of course, follow all the safe computer upgrade guidelines I provided earlier in this chapter.

Improving your Mac's memory

Digital video editing uses a lot of computer memory, so you should see significant performance improvements if you upgrade your RAM. Memory is usually pretty easy to upgrade in most desktop Macs. In fact, most iMacs incorporate a handy access panel that enables you to add memory in mere seconds. Memory comes on little cards called *DIMMs* (dual inline memory modules) and they easily snap into place in special memory slots on the computer's motherboard. Figure 2-1 illustrates what a DIMM looks like. Read the documentation from Apple that came with your Mac for specific instructions on installing more memory.

Make sure you obtain memory that is specifically designed for your Mac — specify the model as well as its processor type and speed (for example: 500MHz G3). Memory modules come in various sizes, so even if all memory slots in your Mac appear full, you might be able to upgrade by replacing your current DIMMs with bigger ones.

Portable Macs can also receive memory upgrades, but the task is more technically challenging. It usually requires that you remove the keyboard and some other parts of your iBook or PowerBook. I recommend that you leave such upgrades to a professional unless you really know exactly what you are doing.

Upgrading Mac hard drives

The hard drive in virtually any Mac can be replaced with a bigger unit. Most modern Macs have EIDE (Enhanced Integrated Drive Electronics) hard drive, a standardized hard drive format that's ubiquitous in the PC world.

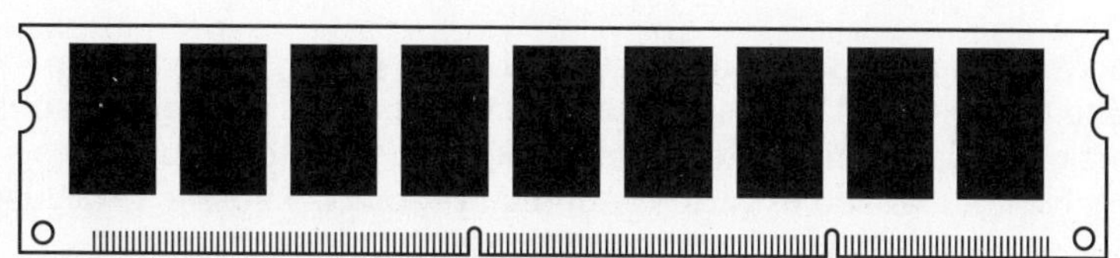
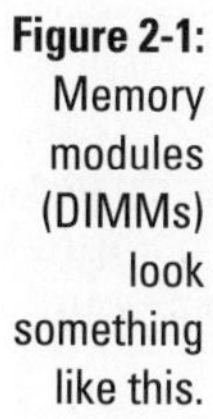

Figure 2-1: Memory modules (DIMMs) look something like this.

Standardization keeps the initial cost of new Macs affordable, and it tends to make replacement parts cheaper and easier to find. But resist the urge to run off and buy (for example) that gargantuan 200GB EIDE drive advertised in a Sunday flyer to replace the 4GB drive in your four-year-old Bondi blue iMac. Hold your horses until you consider two potential problems:

- **The new hard drive may be too big.** Older Macs may not support some of the massive newer hard drives available today.
- **The new hard drive may be too hot.** Literally. One of the greatest challenges facing computer hardware engineers today involves heat management. Modern processors and hard drives put off a lot of heat, and if too much heat is allowed to build up inside the computer's case, the life of your Mac will be greatly shortened. Many older iMacs were not designed to manage the heat generated by the newest, fastest hard drives.

You should only buy a hard drive recommended for your specific Macintosh model. This means you're best off buying a new hard drive from a knowledgeable Macintosh retailer. As with memory, make sure you specify your Mac's model and processor when purchasing a hard drive.

Adding an external hard drive to your Mac

By far, the easiest way to add storage space to your Mac is to use an external hard drive. External drives that connect to a FireWire or USB port are widely available, and although they tend to be more expensive than internal drives, their ease of installation and use makes them worthwhile.

Unfortunately, external drives are usually less than ideal for working with digital video. Even a FireWire or USB 2.0 (a newer, faster USB standard) external drive will be slower than an internal drive — and drive speed is crucial when you are capturing or exporting video. External hard drives are fine if you need a big place to store music or other files, but I don't recommend using an external drive as your main drive for video work.

Choosing Mac video software

Apple offers a pretty good selection of video-editing software for the Macintosh — good thing they do, too, because not many other software vendors offer Mac-compatible editing programs. Chapter 19 provides a comprehensive comparison of various editing programs; if you're a Mac user, your choices are pretty much limited to these:

- **Adobe Premiere:** Available for both Windows and Mac, this was one of the first pro-caliber video-editing programs for personal computers. It's a little expensive (about $550), but provides an advanced, power set of video-editing tools. For more on this great program, check out *Adobe Premiere For Dummies* by yours truly, published by Wiley Publishing, Inc..
- **Apple iMovie:** It comes free with all new Macintosh computers, and you can download the latest version of iMovie for free from `www.apple.com`. iMovie 3 is featured throughout this book.
- **Apple Final Cut Pro:** If you *can* afford pro-level prices and you want one of the most cutting edge video-editing programs available, consider Final Cut Pro (about $1000). Final Cut Pro, shown in Figure 2-2, is used by many professional video editors.

Figure 2-2: Many video professionals do their work using Final Cut Pro on a Macintosh.

- **Apple Final Cut Express:** This program offers many of the features of Final Cut Pro for a fraction of the price (about $300). Final Cut Express is a good choice if you want pro-style editing features but can't afford pro-level prices.
- **Avid Xpress DV:** Avid has been making professional video-editing workstations and gear for years, so it's no surprise that they also offer one of the most advanced video-editing programs as well. And it better be good too, because it retails for $1699 (for Windows or Mac).

As you can see, if you're a professional video editor, you can choose between several programs for your Mac. If you're not a pro, however, your choices are a little more limited. Fortunately, iMovie is a reasonably capable program. You can even expand the capabilities of iMovie with plug-ins from Apple (and some third parties). Visit `www.apple.com/imovie/` and check out the iMovie Downloads section for more information.

Using a Windows PC

Macintosh computers have long been favored by professional graphic and video artists, but you can do some pretty advanced video editing on a Windows PC as well. In fact, the greater variety of editing software available for Windows means that these days you can do just about anything on a PC that you can do on a Mac. The next few sections show you what to consider when choosing and preparing a Windows PC for digital video.

Buying a new PC

Countless PC vendors offer computers running Windows for just about any budget these days. For about the same amount of money as you would have spent just to buy a printer 10 years ago, you can now buy a new PC — including monitor — and if you shop around, you might even find someone to throw in the printer for free.

You may find, however, that a bargain-basement computer is not quite good enough for digital video. The hard drive may be too small, the processor may not be fast enough, the computer might not have enough memory, or some other features may not be ideal. Look for these features when you're shopping for a new PC:

- **Windows XP:** Some new PCs might come with Windows Me (Millennium Edition). Windows Me has some fundamental problems with stability and memory management that (in my opinion) make it unsuitable for digital video work. Windows XP, on the other hand, is very efficient and stable. Upgrading a Windows Me machine to XP is often challenging, so I recommend that you buy a PC that already has XP installed.

- **512MB RAM:** Video editing requires a lot of random-access memory (RAM) — the more the better.

As if you didn't already have enough acronyms to remember, some PCs have a type of memory called DDR (Double Data Rate) RAM. DDR works twice as efficiently as regular RAM, so a computer with 256MB of DDR RAM will work about as well as a computer with 512MB of other types of RAM.

- **32MB video RAM:** The video image on your monitor is generated by a component in your computer called the *video card* or *display adapter*. The video card has its own memory — I recommend at least 32MB. Some video cards share system RAM (the computer's spec sheet might say something like "integrated" or "shared" in reference to video RAM). This tends to slow down the performance of your computer, so I recommend that you avoid shared video RAM.
- **1GHz (gigahertz) processor:** I recommend a processor speed of at least 1 GHz (equal to 1000MHz) or faster. This shouldn't be a problem because there aren't too many PCs still being sold with processors slower than 1 GHz. It really doesn't matter if the processor is an Intel Pentium, an AMD Athlon, or even an AMD Duron. The faster the better, naturally.
- **FireWire:** Unlike Macs, not all PCs come with FireWire (also called IEEE-1394) ports. You can upgrade most PCs with a FireWire card, but buying a computer that already has FireWire is a lot easier.
- **60GB hard drive:** When it comes to hard drives, bigger is better. If you plan to do a lot of video-editing work, 60GB is an absolute minimum. Sure, it *sounds* like a lot, but you'll use it up in a hurry as you work with digital video.

Another option, of course, is to build your own computer. For some, the act of building a PC from scratch remains a vaunted geek tradition (you know who you are). If you choose to build your own, make sure that your system meets — preferably *exceeds* — the guidelines given here. I built a computer last year tailored specifically for video editing — and it only cost me about $400 plus some spare parts scrounged from my own stocks. But know what you're doing before you start down this path; it's definitely not the path of least resistance. Heed the wisdom of this ancient computer-geek proverb: *Building your own PC is cheap only if your time is worthless!*

If you're a Linux devotee, good for you! However, if you plan to do much with digital video, you're going to have to bite the proverbial bullet and use either a Mac or the dreaded Windoze. Currently the mainstream offers no video-editing programs designed for Linux. Although some tools will allow you to run some Windows or Mac programs, system performance often deteriorates, meaning you probably won't be able to capture and edit video efficiently.

Upgrading your PC for digital video

If you already have a PC, you can probably do some things to make it better suited for video work. These tweaks may or may not be simple, however. The diversity of the PC market means your computer can probably be upgraded, but performing hardware upgrades requires experience and expertise (whether your own or somebody else's).

When considering any upgrade, your first step is to check your PC's documentation for information about what upgrades can be performed. The manufacturer may offer upgrade kits designed specifically for your computer. The following sections address some specific upgrades you may be considering. Of course, make sure you follow the computer-upgrade guidelines I provide earlier in this chapter. I also recommend buying a book that specifically covers PC upgrades, such as *Upgrading and Fixing PCs For Dummies,* by Andy Rathbone, published by Wiley Publishing, Inc..

Choosing a new hard drive

Not so long ago, I bought a 1.6GB hard drive for *just* $200. At the time, I couldn't imagine using up all that space, and the price seemed like an almost unbelievably good deal. But drive sizes have grown and prices have dropped — fast enough to make a lot of heads spin. Nowadays that $200 will buy you a 160GB hard drive — a hundred times more space — and that's if you don't shop around for a better deal.

Big, cheap hard drives are a popular PC upgrade these days. They have become so ubiquitous that even one of my local grocery stores sells hard drives. Digital video consumes hard drive space almost as fast as my sons gobble chips and salsa at the local Tex-Mex restaurant, so there's a good chance that you may be considering a hard drive upgrade of your own. If so, here are some important considerations before you nab that hard drive along with the broccoli at your local supermarket:

- **EIDE interface:** Your hard drive connects to the rest of the computer through a special disk interface. Most modern PCs use the EIDE (Enhanced Integrated Drive Electronics) interface — and such drives are both fast and widely available. Check your computer's documentation to make sure it uses this type of interface. A few computers use the SCSI (Small Computer System Interface) format, which is also fast, but SCSI drives are expensive and increasingly hard to find. If the PC is older and just uses a regular IDE interface, it probably won't be fast enough to work with digital video; it also may not support the large size of modern hard drives.

- **Drive speed:** EIDE drives are commonly available in speeds of 5400 rpm or 7200 rpm. This speed is always clearly marked on the drive's packaging. For digital video work, a 7200-rpm drive is virtually mandatory.
- **Windows installation:** If you're replacing your old hard drive with a newer, bigger drive, you'll have to (you guessed it) reinstall Windows. Do you have a Windows installation CD? You should have received an installation disc with your PC. If not, figure on spending at least $199 for a new Windows CD (you'll need to buy the Full Version, not a cheaper Upgrade Version).
- **Support for big hard drives:** Maybe your computer won't support the biggest hard drives sold today (admittedly a remote possibility). Check the BIOS section of your PC's documentation to see whether your computer has any drive-size limitations.

If you do decide to upgrade to a bigger hard drive, you can either replace your current hard drive, or (in some cases) simply add a second hard drive to your computer. If you look inside your computer case, you'll probably see a gray ribbon cable running from the motherboard to the hard drive. In some systems, that cable has two connectors — one connecting to the hard drive and the other to the CD-ROM drive. This arrangement is called *daisy-chaining*, and sometimes you can get away with daisy-chaining two hard drives together. If you can, it's really cool — you just keep your applications where they are and add a second great big hard drive to use exclusively as a scratch pad for all your video work. Again, consult your PC's documentation (and a book like *Upgrading and Fixing PCs For Dummies*) to see whether daisy-chaining is an option.

Finally, as I mentioned in the Macintosh section of this chapter, external hard drives (even FireWire drives) usually aren't fast enough for video work. I recommend you stick with internal drives for the bulk of your video work.

Upgrading to Windows XP

Earlier I recommended that you run Windows XP if you plan to work with digital video. Windows XP is vastly more stable than previous versions of Windows (especially Windows 95, 98, and Me) — and it does a much better job of managing system memory, which is crucial when you work with video.

If you don't already have Windows XP, you'll need to upgrade. If you already have a version of Windows on your computer, you can purchase an upgrade to Windows XP Home Edition for $99 or Windows XP Professional for $199. If you don't already have Windows, the Full Version of XP Home will set you back $199, and the full version of XP Pro costs $299. Windows XP Professional is nice, but so is XP Home. If you don't need the extra networking tools built into Windows XP Pro, XP Home is just fine for video work.

You may have heard horror stories from friends who tried to upgrade their old computers to Windows XP. To be honest, I have a horror story of my own (but I'll spare you the gruesome details). For now, I can offer these three pieces of advice:

- **Avoid installing Windows XP on any computer that is more than two or three years old.** XP is a snob about modernity, and may not support some of your older components and hardware. A quick way to check the hardware in your computer is to use Microsoft's online Upgrade Advisor. Visit `www.microsoft.com/windowsxp/pro/howtobuy/upgrading/advisor.asp` for instructions on how to download and use the Upgrade Advisor, which inspects your system and advises whether or not your computer is ready for Windows XP.
- **Perform a "clean" installation.** Although the installation CD provides an option to upgrade your current version of Windows, there is a really, really, *really* good chance that this approach will cause you troubles in the near future. So back up all your important data *before you begin installing,* and let the installer program reformat and repartition your hard drive using the NTFS (NT File System) when you are presented with these options. The rest of the installation process is pretty simple. (Oh, yeah, you'll have to restore all the data from your backup onto that reformatted hard drive. You could be at it for a while.)
- **Check for online updates immediately after installation.** Microsoft is constantly developing updates and fixes for Windows XP, and you can quickly download and install those updates by choosing Start ⇨ All Programs ⇨ Windows Update. You'll have to connect to the Internet to do so, so make sure you have all the necessary information handy to reinstall your Internet service.

Installing a FireWire card

If you have a PC and want to work with digital video, perhaps the one upgrade you are most likely to perform is to install a FireWire (IEEE-1394) card. A FireWire card is crucial if you want to capture video from a digital camcorder into your computer.

To install a FireWire card, you need to have an empty PCI slot inside your computer. PCI slots are usually white and look like the open slot in Figure 2-3.

If you have an empty PCI slot, you should be able to install a FireWire card. Numerous cards are available for less than $100. Most FireWire cards also come packaged with video-editing software, so consider the value of that software when you make your buying decision. Pinnacle (`www.pinnaclesys.com`), for example, sells Pinnacle Studio DV — which includes both a FireWire card and a full version of the Studio editing software (featured throughout this book, by the way) for a retail price of $129. Shop around and you may find an even better sale price!

Figure 2-3: Make sure you have an empty PCI slot to accommodate a new FireWire card.

When you purchase a FireWire card, read the box to make sure your computer meets the system requirements. Follow the installation instructions that come with the card to install and configure your FireWire card for use. After the card is installed, Windows automatically detects your digital camcorder when you connect it to a FireWire port and turn on the camcorder's power. You can then use Windows Movie Maker or other video-editing software to capture video from the camcorder.

Choosing Windows video software

Perhaps the nicest thing about using a Windows PC is that no matter what you want to do, lots of software is available to help you. Countless video-editing programs are available, and in Chapter 19, I provide a feature-specific cross reference of several popular editing programs. Video-editing software for Windows breaks down into three basic categories:

- **Basic:** At the low end of the price-and-feature scale are free programs (such as Windows Movie Maker) or programs that come free with cheaper FireWire cards (such as Ulead VideoStudio or Roxio VideoWave). These programs are usually pretty limited in terms of what you can do with them; I recommend moving up to the next level as soon as your budget allows.

- **Intermediate:** A growing number of video-editing programs now offer more advanced editing features at a price that is not out of the average consumer's reach. Pinnacle Studio — featured throughout this book — is one good example. Studio retails for $99 and offers more advanced editing than most basic programs — plus you can expand the capabilities even more with the Hollywood FX Plus plug-in ($49).
- **Advanced:** If you're willing to spend $400 or more, you can get some of the same programs that the video pros use. Advanced video-editing programs include Adobe Premiere, Avid Xpress DV, Pinnacle Edition, and Sonic Foundry Vegas.

The software you choose depends greatly upon your budget and your needs. Again, check out Chapter 19 for a complete feature-and-price comparison of some top video-editing programs currently available for your PC.

Optimizing Windows for video work

Even if you have a brand new computer with a wicked-fast processor and lots of RAM, you may experience problems when you work with digital video. Perhaps the most common problem is dropped frames during capture or export. A *dropped frame* occurs when your computer can't keep up with the capture or export process and loses one or more video frames. Pinnacle Studio and most other video-editing programs report dropped frames if they occur. If you encounter dropped frames (or you just want to help your computer run more efficiently for video work), try the tips in the following sections to improve performance.

I recommend running Windows XP for digital video work, and the following sections assume you are using XP (whether the Pro or Home version). If you're using an earlier version of Windows (such as Me), you can still follow along, though some steps may be slightly different.

Updating video drivers

Windows operates the various components in your computer by using software tools called *drivers*. Outdated drivers can cause your computer to run slowly — or even crash. This is especially true of video display drivers in Windows XP. The *display adapter* (another name for the video card) is the component that generates the video image for your computer's monitor. Hardware vendors frequently provide updates, so check the manufacturers' Web sites regularly for downloadable updates. If you aren't sure who made your display adapter, follow these steps:

1. **Choose Start➪Control Panel.**
2. **In the Windows Control Panel, click Performance and Maintenance if you see that option, and then double-click the System icon.**

   If you do not see a Performance and Maintenance listing, simply double-click the System icon.

   The System Properties dialog box appears.
3. **Click the Hardware tab to bring it to the front, and then click Device Manager.**

   The Windows Device Manager opens.
4. **Click the plus sign next to Display Adapter.**

   More information about the display adapter appears in the list, including the manufacturer and model of your display adapter.
5. **Click the Close (X) button when you are done to close the Device Manager.**

As you can see in Figure 2-4, my display adapter is made by NVIDIA. To check for updated drivers, I can do a Web search for the manufacturer's name, and then visit their Web site to see if updates are available. The Web site should contain installation instructions for the driver updates. Make sure that any driver updates you download are designed specifically for Windows XP (or specifically for the version of Windows you're using).

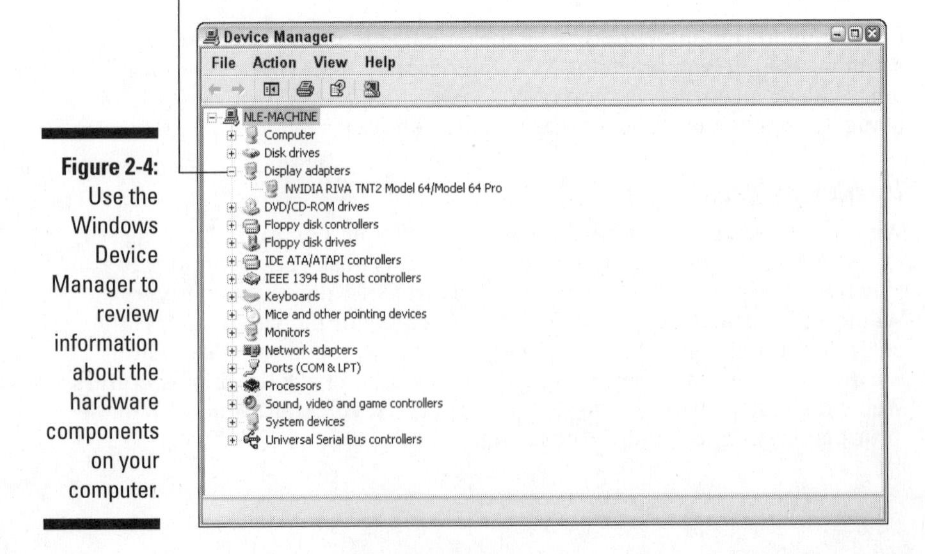

Figure 2-4: Use the Windows Device Manager to review information about the hardware components on your computer.

Adjusting power settings and screen savers

Your computer is probably set to power down after a period of inactivity. Normally this is a good thing because it conserves energy, but it can cause problems during video capture and some other video-editing actions. To temporarily disable power-saving options, follow these steps:

1. **Choose Start⇨Control Panel.**
2. **In the Windows Control Panel, click Performance and Maintenance if you see that option, and then double-click the Power Options icon.**

   If you do not see a Performance and Maintenance listing, simply double-click the Power Options icon.

   The Power Options Properties dialog box appears.
3. **On the Power Schemes tab, set all four pull-down menus (near the bottom of the dialog box) to Never, as shown in Figure 2-5.**

   Alternatively, you can just choose the Always On scheme from the Power Schemes pull-down menu.
4. **Click Save As, name the power scheme *Video* in the Save Scheme dialog box that appears, and then click OK.**
5. **Click OK again to apply the changes and close the Power Options Properties dialog box.**

If you want to conserve power in the future, you can turn on the power-saving features by simply choosing a different power scheme in the Power Options Properties dialog box.

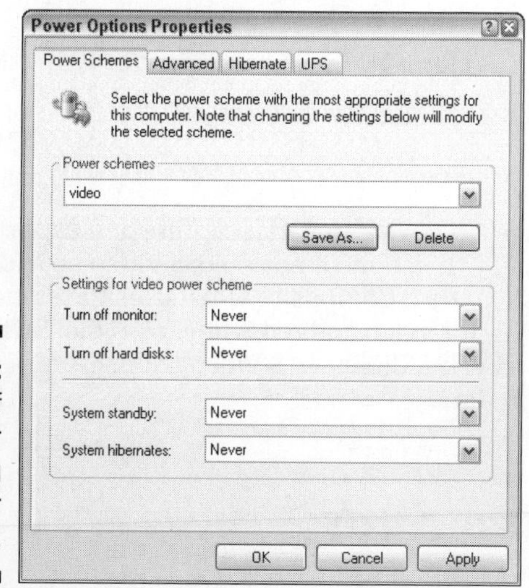

Figure 2-5: Turn off power saving features for video work.

I also recommend that you disable screen savers when you are getting ready to work with video. To do so, right-click an empty area of the Windows desktop and choose Properties. In the Display Properties dialog box, click the Screen Saver tab and choose (None) in the Screen saver menu. Click OK to close the Display Properties dialog box.

Choosing Analog Capture Hardware

If you want to capture video from a digital camcorder, the best way to do so is with a FireWire port. But if you want to capture analog video — whether from a VCR, Hi8 camcorder, or other analog source — you'll need some specialized hardware. You can install a video-capture card in your computer, or (possibly) use an external *analog video converter* that connects the analog device to your computer's FireWire or USB port.

Read the packaging carefully before you buy any video-capture hardware and make sure that it is designed to capture analog hardware. Some FireWire cards are marketed simply as video-capture cards, even though they can only capture *digital* video.

When choosing an analog video-capture device, check the packaging to make sure your computer meets the system requirements. The device should also be capable of capturing the following:

- NTSC (North America, Japan, the Philippines) or PAL (Australia, South America, Southeast Asia, most of Europe) video, whichever matches your local video standard (see Chapter 3 for more on broadcast video standards)
- 30 frames per second (fps) for NTSC video or 25 fps for PAL video
- 720 x 534 (NTSC) or 768 x 576 (PAL) video frames
- Stereo audio

Although you don't have to get a device that can capture and export S-Video as well as composite video, try to get one if possible; S-Video provides better image quality than composite video. Composite video uses the standard RCA-style jacks, color-coded yellow for video and red/white for audio; S-Video connectors look like Figure 2-6. S-VHS VCRs have S-Video connectors, as do some higher-quality analog camcorders.

Selecting capture cards

Earlier in this chapter, I provide an overview of how to install a FireWire card in a Windows PC. Analog video-capture cards are usually installed inside your computer in much the same way. That means you'll have to have some expertise in upgrading computer hardware, and you should follow the computer-upgrade guidelines I detail in that earlier section of the chapter.

Figure 2-6: Try to get a device with an S-Video connector like this.

Okay, it may seem obvious, but I'll say it anyway: Make sure your computer has an available expansion slot of the correct type in which you can install the card.

Many capture cards have neat little accessories called *breakout boxes*. The space available on the back of an expansion card is pretty small, and may not provide enough room for all the needed audio and video ports. Instead, the ports will reside in a breakout box — which can sit conveniently on your desk. The breakout box connects to the capture card using a special (included) cable. Figure 2-7 shows the card and breakout box that you get with Pinnacle Studio Deluxe. This product allows both digital and analog video capture because it not only has *two* FireWire ports on the card, but also S-Video and composite video ports on a breakout box.

Selecting external video converters

If you don't feel like ripping into the innards of your computer, you may want to consider an external analog video converter, such as the Dazzle Hollywood DV Bridge. These devices usually connect to your computer's FireWire or USB (Universal Serial Bus) port. You connect your VCR or analog camcorder to the converter, connect the converter to your computer, and the analog video is converted into digital video as it is captured into your computer. Chapter 18 gives more information on a couple of popular video converters.

If you buy a USB converter, make sure that both the device and your computer use USB 2.0 (a newer, faster version of USB). The original version of USB could only transfer data at 12Mbps (megabits per second), which is not quite enough for full-quality video capture. USB 2.0, however, can transfer 480Mbps, which is even faster than FireWire.

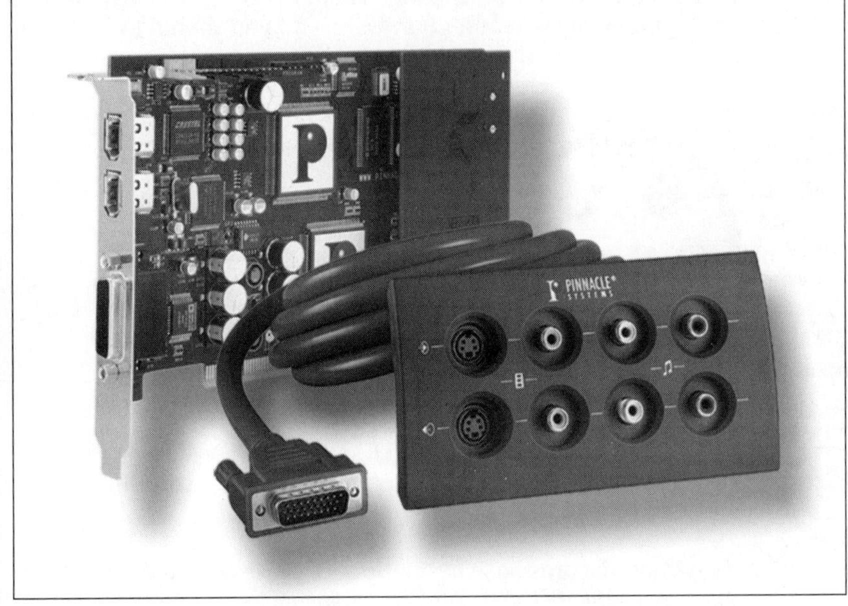

Figure 2-7: Some capture cards put all their audio and video connectors on a breakout box.

© Pinnacle Systems, Inc.

Improving the Human-Machine Interface

A lot of science goes into the ergonomics of computers. Modern keyboards, monitors, and mice are well designed, and make using your computer both pleasant and healthful — but they're designed for *general* computer use. Working with video is a lot easier if you have some specialized gear to improve the interface between you and your computer. The following sections suggest two special items you may want to use to make video work easier.

Working with video monitors

Computer monitors and TV screens may look similar, but the two have profound technological differences. The most important difference involves color. Computer monitors can display more colors than TV screens. Also, computer screens are non-interlaced; TVs are usually interlaced (interlaced displays draw every other line of the picture on separate passes, whereas a non-interlaced or progressive scan display draws the whole picture at once). Chapter 3 goes into more detail about video color and interlacing, but for now the important point is that the video you preview on your computer monitor may look a lot different when it's viewed on a TV.

What is an editing appliance and do I need one?

Several companies market versions of an expensive gadget called an *editing appliance* — basically a computer designed to do nothing but edit video. It has a hard drive, capture ports, and built-in editing software. Editing appliances are usually quite powerful — and so is the price. The Casablanca Prestige from Macro System (`www.casablanca.tv`) carries a suggested retail price of $3495 (monitor not included).

Unless you are a professional video editor working in a broadcast environment (in which case you probably aren't reading this book), you really don't need an editing appliance. A regular computer (PC or Mac) provides you with a lot more value for your money because not only can you do some pretty advanced video editing on your computer, you can also send e-mail, type memos for work, and shop on eBay. And when your computer is obsolete in a few years, a local school or youth organization will be happy to take it as a donation. An obsolete editing appliance, on the other hand, won't even make a good boat anchor five years from now.

To address this problem, many video editors connect a video monitor (that is, a TV) to their computers so they can preview how the video looks on a real TV. Fortunately, you don't need expensive, specialized hardware to hook up a video monitor to your computer. All you need is an old color TV and one of the following devices that connect to your computer:

- **An analog capture device:** If you have an analog capture card or video converter, you might be able to hook up a monitor to the video output connectors. Check the capture device's documentation for instructions on connecting a video monitor.
- **A digital camcorder:** Connect a TV to the analog outputs on your digital camcorder, and connect the camcorder to your FireWire port. You can even use the camcorder itself as a monitor — but keep in mind that the LCD display on your camcorder is probably non-interlaced, so you won't be seeing a "real" TV picture. (Is "real TV" an oxymoron? Who knows? Just make sure your movie looks right on an interlaced screen.)

Some video-editing programs allow you to play video directly to an external monitor. In Pinnacle Studio, you must first export the movie as if you were going to export it to tape. After you have exported a file, simply connect your monitor and click Play in Studio's preview window. See Chapter 15 for complete instructions on how to export your movie to tape.

If you have titles or other graphics in your movie that incorporate very thin lines, interlacing could cause the graphics or letters to flicker when they're viewed on a TV. Pay special attention to anything with very thin lines when you preview your movie on a video monitor. Chapter 9 provides tips on making better titles.

Using a multimedia controller

A lot of video editing involves finding exactly the right spot to make a cut or insert a clip. The ability to easily move back and forth through video precisely, frame-by-frame, is crucial, but it's also not terribly easy when you are using the keyboard and mouse. For years, professional video-editing workstations have used knobs and dials to give editors more intuitive, precise control — and now you can get that same level of control on your computer. A *multimedia controller* such as the SpaceShuttle A/V from Contour Design (`www.contouravs.com`) connects to your computer's USB port and makes manipulating video a lot easier. The SpaceShuttle A/V is shown in Figure 2-8.

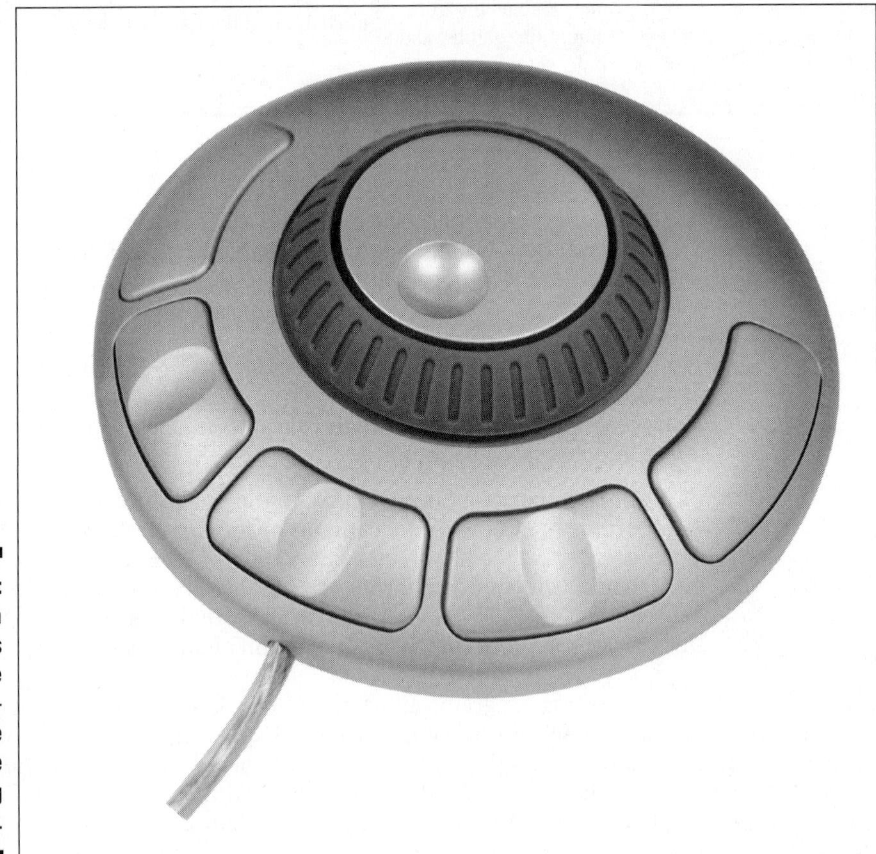

Figure 2-8: Multimedia controllers like the Space-Shuttle A/V make video editing a lot easier.

Photo courtesy of Contour A/V Solutions, Inc.

I have used the ShuttlePRO (another controller from Contour Design) extensively with Adobe Premiere, Apple iMovie, Final Cut Pro, and Pinnacle Studio, and it truly makes common editing tasks a breeze. I don't have to spend time trying to remember which keyboard key starts playback or moves to the next frame; instead, I just use the simple, intuitive controls on the multimedia controller to swiftly and effortlessly control my editing program. (For more on multimedia controllers, see Chapter 19.)

Chapter 3

Getting Your Digital Video Gear

In This Chapter

- Selecting a digital camcorder
- Choosing audio equipment
- Shedding light on your video shoots
- Tripods

If you have just turned here from Chapter 2 hoping to find out how to achieve world peace, I am sorry, but I cannot be of much help. My editors noted that world peace was not in the original outline, so that content had to be cut. Also, there was some question as to whether I successfully resolved that Mac-versus-PC thing. (By the way, if your name is Jimmy Carter and you're interested in writing *World Peace For Dummies*, please give the publisher a call.)

Fortunately, I *can* tell you about digital video gear because it was approved in the outline. The most important piece of gear is, of course, a digital camcorder. I'll also help you pick out audio gear, lighting, tripods, and other gear that will help you make better movies.

Choosing a Camcorder

Digital camcorders — also called *DV* (digital video) camcorders — are among *the* hot consumer electronics products today. This means that you can choose from many different makes and models, with cameras to fit virtually any budget. But cost isn't the only important factor as you try to figure out which camera is best for you. You need to read and understand the spec sheet for each camera and determine if it will fit your needs. The next few sections help you understand the basic mechanics of how a camera works, as well as compare the different types of cameras available.

I won't make specific camera model recommendations — the market is constantly changing — but I can list some up-to-date resources to help you compare the latest and greatest digital camcorders. My favorites are

- **CNET.com (`www.cnet.com`):** This is a great online resource for information on various computer and electronics products. The editorial reviews are helpful, and you can read comments from actual owners of the products being reviewed. The Web site also provides links to various online retailers. If you order a camcorder online, consider more than just the price. Find out how much shipping will cost, and pay close attention to the retailer's rating on CNET. Retailers earn high star ratings by being honest with customers and fulfilling orders when promised.
- **The local magazine rack:** Visit Barnes & Noble, Borders, Waldenbooks, or any other bookstore that has a good magazine selection. There you should find magazines and buyer's guides tailored to you, the digital video enthusiast. *Computer Videomaker* is one of my favorites, although many of its articles are aimed at professional and semi-pro videographers.

Mastering Video Fundamentals

Before you go in quest of a camcorder, it's worth reviewing the fundamentals of video and how camcorders work. In modern camcorders, an image is captured by a *charged coupled device* (CCD) — a sort of electronic eye. The image is converted into digital data and then that data is recorded magnetically on tape.

The mechanics of video recording

It is the springtime of love as John and Marsha bound towards each other across the blossoming meadow. The lovers' adoring eyes meet as they race to each other, arms raised in anticipation of a passionate embrace. Suddenly, John is distracted by a ringing cell phone and he stumbles, sliding face-first into the grass and flowers at Marsha's feet. A cloud of pollen flutters away on the gentle breeze, irritating Marsha's allergies, which erupt in a massive sneezing attack.

As this scene unfolds, light photons bounce off John, Marsha, the blossoming meadow, the flying dust from John's mishap, and everything else in the shot. Some of those photons pass through the lens of your camcorder. The lens focuses the photons on transistors in the CCD. The transistors are excited, and the CCD converts this excitement into data, which is then magnetically recorded on tape for later playback and editing. This process, illustrated in Figure 3-1, is repeated approximately 30 times per second.

Most mass-market DV camcorders have a single CCD, but higher-quality cameras have three CCDs. In such cameras, individual CCDs capture red, green, and blue light, respectively. Multi-CCD cameras are expensive (typically over $1500), but the image produced is near-broadcast quality.

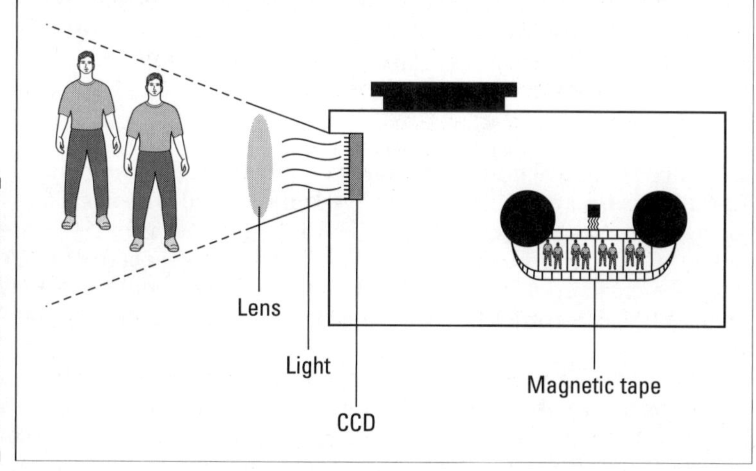

Figure 3-1: The CCD converts light into the video image recorded on tape.

Early video cameras used video pickup tubes instead of CCDs. Tubes were inferior to CCDs in many ways, particularly in the way they handled extremes of light. Points of bright light (such as a light bulb) bled and streaked light across the picture, and low-light situations were simply too dark to shoot.

Broadcast formats

A lot of new terms have entered the videophile's lexicon in recent years: NTSC, PAL, SECAM. These terms identify broadcast television standards, which are vitally important to you if you plan to edit video — because your cameras, TVs, tape decks, and DVD players probably conform to only one broadcast standard. Which standard is for you? That depends mainly on where you live:

- **NTSC (National Television Standards Committee):** Used primarily in North America, Japan, and the Philippines.
- **PAL (Phase Alternating Line)** Used primarily in Western Europe, Australia, Southeast Asia, and South America.

- **SECAM (Sequential Couleur Avec Memoire)** This category covers several similar standards used primarily in France, Russia, Eastern Europe, and Central Asia.

What you need to know about video standards

The most important thing to know about the three broadcast standards is that they are *not* compatible with each other. If you try to play an NTSC-format videotape in a PAL video deck (for example), the tape won't work, even if both decks use VHS tapes. This is because VHS is merely a physical *tape* format, and not a *video* format.

On a more practical note, make sure you buy the right kind of equipment. Usually this isn't a problem. If you live in the United States or Canada, your local electronics stores will only sell NTSC equipment. But if you are shopping online and find a store in the United Kingdom that seems to offer a really great deal on a camcorder, beware: That UK store probably only sells PAL equipment. A PAL camcorder will be virtually useless if all your TVs are NTSC.

Another consideration involves video-editing software. Many video-editing programs are compatible with both NTSC and PAL video, so make sure to use program settings that match whatever video format you are working with. To check settings in the programs featured in this book:

- **Apple iMovie:** iMovie automatically detects your standard when you capture video. If you want to export to a different format, choose File➪Export, select To QuickTime in the Export menu, and choose Expert Settings in the Formats menu. Click Export, and then click Options in the Save Exported File As dialog box. Click Settings under Video, and choose a compressor that matches the standard you want to use (such as DV–NTSC or DV–PAL).
- **Pinnacle Studio:** Open Studio and choose Setup➪Capture Source. Make sure that the TV Standard menu is set to your local standard.
- **Windows Movie Maker:** From within the program, choose Tools➪Options. On the Advanced tab of the Options dialog box, choose NTSC or PAL under Video Properties.

About 99.975% of the time, you won't need to change your video standard. You should only adjust video standard settings if you know that you are working with video from one standard and need to export it to a VCR or camcorder that uses a different standard.

Some nice-to-know stuff about video standards

Video standards differ in two primary ways. First, they have different frame rates. The *frame rate* of video is the number of individual images that appear per second, thus providing the illusion that subjects on-screen are moving. Frame rate is usually abbreviated *fps* (frames per second). Second, the standards use different resolutions. Table 3-1 details the differences.

You're probably familiar with resolution measurements for computer monitors. Computer resolution is measured in pixels. If your monitor's resolution is set to 1024 x 768, the display image is 1024 pixels wide by 768 pixels high. TV resolution, on the other hand, is usually measured by the number of horizontal lines in the image. An NTSC video image, for example, has 525 lines of resolution. This means that the video image is drawn in 525 separate horizontal lines across the screen.

Table 3-1 Video Standards

Standard	*Frame rate*	*Resolution*
NTSC	29.97 fps	525 lines
PAL	25 fps	625 lines
SECAM	25 fps	625 lines

Interlacing versus progressive scan

A video picture is usually drawn as a series of horizontal lines. An electron gun at the back of the picture tube draws lines of the video picture back and forth, much like the printer head on your printer moves back and forth as it prints words on a page. All three broadcast video standards — NTSC, PAL, and SECAM — are *interlaced*; the horizontal lines are drawn in two passes rather than one. Every other line is drawn on each consecutive pass, and each of these passes is called a *field*. So on a PAL display, which shows 25 fps, there are actually 50 fields per second.

Noninterlaced displays are becoming more common. Modern computer monitors, for example, are all *noninterlaced* — all the lines are drawn in a single pass. Some HDTV (high-definition television) formats are noninterlaced; others are interlaced.

Most camcorders record interlaced video, but some high-end camcorders also offer a progressive-scan mode. *Progressive scan* — a fancy way of saying noninterlaced — is a worthwhile feature if you can afford it, especially if you shoot a lot of video of fast-moving subjects. With an interlaced camera, a problem I call *interlacing jaggies* can show up on video with a lot of movement. Figure 3-2, for example, shows video of a fast-moving subject that was shot using an interlaced camera. The stick seems to have some horizontal slices taken out of it, an unfortunate illusion created because the subject was moving so fast that the stick was in different places when each field of the video frame was captured. If the entire frame had been captured in one pass (as would have happened with a progressive-scan camcorder), the horizontal anomalies would not appear.

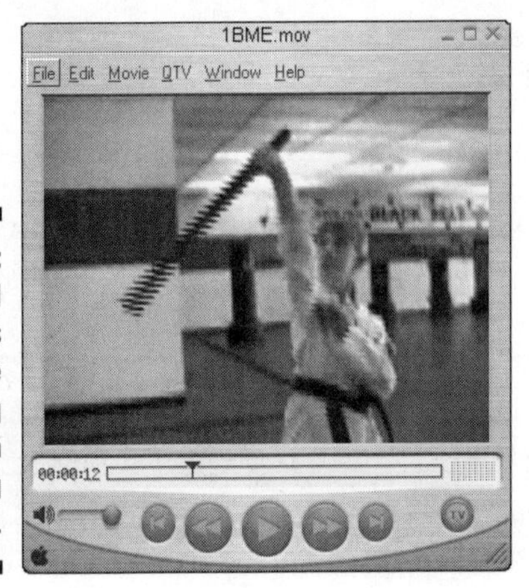

Figure 3-2: Interlaced cameras can cause interlacing jaggies on fast-moving subjects.

The many aspects of aspect ratios

Different moving picture displays have different shapes. The screens in movie theaters, for example, look like long rectangles; most TV screens and computer monitors are almost square. The shape of a video display is called the *aspect ratio*. The following two sections look at how aspect ratios effect your video work.

Image aspect ratios

The aspect ratio of a typical television screen is 4:3 (four to three) — for any given size, the display is four units wide and three units high. To put this in real numbers, measure the width and height of a TV or computer monitor that you have nearby. If the display is 32 cm wide, for example, you should notice that it's also about 24 cm high. If a picture completely fills this display, the picture has a 4:3 aspect ratio.

Different numbers are sometimes used to describe the same aspect ratio. Basically, some people who make the packaging for movies and videos get carried away with their calculators, so rather than call an aspect ratio 4:3, they divide each number by three and call it 1.33:1 instead. Likewise, sometimes the aspect ratio 16:9 is divided by nine to give the more cryptic-looking number 1.78:1. Mathematically, these are just different numbers that mean the same thing.

A lot of movies are distributed on tape and DVD today in *widescreen* format. The aspect ratio of a widescreen picture is often (but not always) 16:9. If you watch a widescreen movie on a 4:3 TV screen, you will see black bars (also called letterboxes) at the top and bottom of the screen. This format is popular because it more closely matches the aspect ratio of the movie-theater screens for which films are usually shot. Figure 3-3 illustrates the difference between the 4:3 and 16:9 aspect ratios.

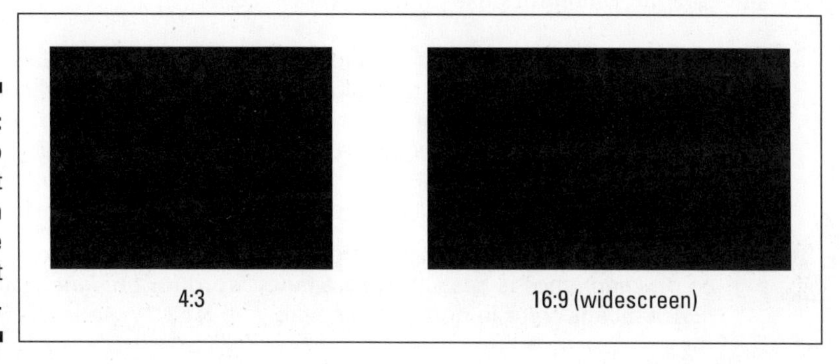

Figure 3-3: The two most common image aspect ratios.

A common misconception is that 16:9 is the aspect ratio of all big-screen movies. In fact, various aspect ratios for film have been used over the years. Many movies have an aspect ratio of over 2:1, meaning that the image is more than twice as wide as it is high! But for many films, 16:9 is considered close enough. More to the point, it's just right for you — because if your camcorder has a widescreen mode, its aspect ratio is probably 16:9.

Pixel aspect ratios

You may already be familiar with image aspect ratios, but did you know that pixels can have various aspect ratios too? If you have ever worked with a drawing or graphics program on a computer, you're probably familiar with pixels. A *pixel* is the smallest piece of a digital image. Thousands — or even millions — of uniquely colored pixels combine in a grid to form an image on a television or computer screen. On computer displays, pixels are square. But in standard video, pixels are rectangular. In NTSC video, pixels are taller than they are wide, and in PAL or SECAM, pixels are wider than they are tall.

Pixel aspect ratios become an issue when you start using still images created as computer graphics — for example, a JPEG photo you took with a digital camera and imported into your computer — in projects that also contain standard video. If you don't prepare the still graphic carefully, it could appear distorted when viewed on a TV. See Chapter 12 for more on using still graphics in your movie projects.

Color

Remember back in the old days when many personal computers used regular televisions for monitors? In the early 1980s I had a Commodore 64 hooked up to a TV — it made sense at the time — but these days it's hard to believe, especially when you consider how dissimilar TVs and computer monitors have become. Differences include interlacing versus progressive scan, horizontal resolution lines, and unique pixel aspect ratios. On top of all that, TVs and computer monitors use different kinds of color.

Computer monitors utilize what is called the *RGB color space*. RGB stands for *red-green-blue*, meaning that all the colors you see on a computer monitor are combined by blending those three colors. TVs, on the other hand, use the *YUV color space*. YUV stands for *luminance-chrominance*. This tells us two things:

- Whoever's in charge of making up video acronyms can't spell.
- Brightness in video displays is treated as a separate component from color. *Luminance* is basically just a fancy word for *brightness*, and *chrominance* means *color* in non-techie speak.

I could go on for pages describing the technicalities of the YUV color space, but there are really only two important things you need to know about color:

- **Some RGB colors won't show up properly on a TV.** This is an issue mainly when you try to use JPEGs or other computer-generated graphics in a video project, or when you adjust the colors of a video image using effects and color settings in your video-editing program. RGB colors that won't appear properly in the YUV color space are often said to be *illegal* or *out of gamut*. You won't get arrested for trying to use them, but they will stubbornly refuse to look right. Generally speaking, illegal colors are ones with RGB values below 20 or above 230. Graphics programs can usually tell you RGB values for the colors in your images. Some graphics programs (like Adobe Photoshop) even have special filters that help you filter out broadcast "illegal" colors from your images.
- **Video colors won't look exactly right when you view them on a computer monitor.** Because you'll probably do most of your video editing while looking at a computer monitor, you won't necessarily see the same colors that appear when the video is viewed on a TV. That's one reason professional video editors connect broadcast-style video monitors to their computer workstations. An external video monitor allows an editor to preview colors as they actually appear on a TV. (See Chapter 13 for more on using a video monitor with your computer.)

Picking a Camera Format

A variety of video recording formats exist to meet almost any budget. By far the most common digital format today is MiniDV, but a few others exist as well. Some digital alternatives are expensive, professionally-oriented formats; other formats are designed to keep costs down or allow the use of very small camcorders. (Various analog formats still exist, though they are quickly disappearing in favor of superior digital formats.) I describe the most common video formats in the following sections.

MiniDV

MiniDV has become the most common format for consumer digital videotape. Virtually all digital camcorders sold today use MiniDV; blank tapes are now easy to find and reasonably affordable. If you're still shopping for a camcorder and are wondering which format is best for all-around use, MiniDV is it.

MiniDV tapes are small — more compact than even audio cassette tapes. Small is good because smaller tape-drive mechanisms mean smaller, lighter camcorders. Tapes come in a variety of lengths, the most common length being 60 minutes.

All MiniDV devices use the IEEE-1394 FireWire interface to connect to computers, and the DV codec serves to compress and capture video. (A *codec* — short for *co*mpressor/*dec*ompressor — is a compression scheme. Chapter 13 tells you more about codecs.) The DV codec is supported by virtually all FireWire hardware and video-editing software.

Digital8

Until recently, MiniDV tapes were expensive and only available at specialty electronics stores, so Sony developed the Digital8 format as an affordable alternative. Digital8 camcorders use Hi8 tapes instead of MiniDV tapes. A 120-minute Hi8 tape can hold 60 minutes of Digital8 video. Initially the cheaper, easily available Hi8 tapes gave Digital8 camcorders a significant cost advantage; however, MiniDV tapes have improved dramatically in price and availability, making the bulkier Digital8 camcorders and tapes less attractive.

Sony still offers a wide variety of affordably priced, high-quality Digital8 camcorders — Hitachi has offered Digital8s as well — and the format has modernized to stay competitive. Digital8 camcorders record digital video using the same DV codec as MiniDV cameras, and Digital8 camcorders also include i.Link (Sony's trade name for FireWire) ports. Digital8 cameras generally offer equivalent video quality to MiniDV cameras of similar price.

If you already have a lot of old Hi8 tapes and you are on a tight budget, a Digital8 camcorder may be worth considering. Digital8 camcorders have an analog mode, which means they can read the analog video recorded by your old Hi8 camcorder.

Keep in mind that Hi8/Digital 8 compatibility only goes one way: Hi8 camcorders cannot read Digital8 video.

Other digital formats

Although MiniDV has become the predominant standard for consumer digital camcorders, many other formats exist. In addition to Digital8 (described in the previous section), the available alternative formats include these:

- **CD-R/W (Compact Disc-Recordable/reWritable):** A few digital camcorders use recordable CDs as the recording medium. The main benefit of this format is that you can place a recorded video CD into any computer with a CD-ROM drive — no FireWire required. Alas, there are downsides: Recording mechanisms are large, the CDs can usually only hold 20 minutes of video, and built-in CD-R/W drives draw a heavy load from camcorder batteries. If you go this route, plan to use a lot of wall current (and a lot of discs). This format has all but disappeared from the marketplace.
- **DVCAM:** Originally developed by Sony for video professionals, this format is based on MiniDV but offers a more robust tape design, higher image quality, and some high-end features designed to appeal mainly to professional video producers and editors. DVCAM camcorders tend to cost about as much as a new economy car, and get much lower gas mileage.
- **DVCPro:** This is another expensive, MiniDV-based, professional-grade format like Sony's DVCAM. Panasonic is responsible for the DVCPro format.
- **Digital Betacam:** Here's yet another professional format that most of us probably can't afford. Digital Betacam (another Sony creation) is based on the dear departed Betacam SP analog format, which for years was a beloved format among professional videographers.
- **MicroMV:** Someone at Sony really likes to create new recording formats. (Remember Betamax?) Sony offers a few consumer-priced camcorders that use the MicroMV format. As the name suggests, MicroMV tapes are really small, allowing MicroMV camcorders to be small and light as well. (Canon somehow manages to make some tiny MiniDV camcorders, but if you must have a teensy-weensy Sony, then MicroMV is your format.)

With any alternative recording format, the first two things you should consider are price and availability of recording media. If you're considering a camcorder that uses the WhizbangDV format, ask yourself how many stores sell WhizbangDV tapes. Will you still be able to find WhizbangDV tapes five years from now?

Analog formats

Analog video has been with us for decades, but it is fading quickly from the scene. Countless analog formats still exist. You've probably seen these formats around, and you might have even owned (or still own) a camcorder that uses one. Besides the generational-loss problems of analog video that I described in Chapter 1, analog formats usually provide fewer horizontal lines of resolution. The very best analog formats offer a maximum of 400 resolution lines — which is where the very cheapest digital formats *start* (lots of MiniDV camcorders offer 500 resolution lines or more). Table 3-2 provides a brief overview of common analog formats.

If you still have an analog camcorder, see Chapter 6 for information on how to capture video from it into your computer.

Table 3-2 Analog Video Formats

Format	*Resolution Lines*	*Description*
VHS	250	Your basic garden-variety videotape; VHS camcorders are bulky
S-VHS	400	A higher-quality incarnation of VHS, but the tapes are still big
VHS-C	250	A compact version of VHS
8mm	260	Smaller tapes mean smaller camcorders
Hi-8	400	A higher-quality version of 8mm
¾ inch Umatic	280	Bulky analog tapes once common in professional analog systems; a higher quality version offers 340 lines
Betacam	300	Sony's professional analog format based on Betamax (remember those?)
Betacam-SP	340	A higher-quality version of Betacam

Choosing a Camera with the Right Features

When you go shopping for a new digital camcorder, you'll be presented with a myriad of specifications and features. Your challenge is to sort through all the hoopla and figure out whether the camera will meet your specific needs. When reviewing the spec sheet for any new camcorder, pay special attention to these items:

- **CCDs:** As mentioned earlier, 3-CCD (also called *3-chip*) camcorders provide much better image quality, but they are also a lot more expensive. A 3-CCD camera is by no means mandatory, but it is nice to have.
- **Progressive scan:** This is another feature that is nice but not absolutely mandatory. (To get a line on whether it's indispensable to your project, you may want to review the section on interlaced video earlier in this chapter.)
- **Resolution:** Some spec sheets list horizontal lines of resolution (for example: 525 lines); others list the number of pixels (for example: 690,000 pixels). Either way, more is better when it comes to resolution.
- **Optical zoom:** Spec sheets usually list optical and digital zoom separately. *Digital zoom* numbers are usually high (200x, for example) and seem appealing. Ignore the big digital zoom number and focus (get it?) on the *optical zoom factor* — it describes how well the camera lens actually sees — and it should be in the 12x-25x range. Digital zoom just crops the picture captured by the CCD and then makes each remaining pixel bigger to fill the screen, resulting in greatly reduced image quality.
- **Tape format:** MiniDV is the most common format, but (as mentioned earlier) for your equipment, using other formats might make more sense.
- **Batteries:** How long does the included battery supposedly last, and how much do extra batteries cost? I recommend you buy a camcorder that uses Lithium Ion batteries — they last longer and are easier to maintain than NiMH (nickel-metal-hydride) batteries.
- **Microphone connector:** For the sake of sound quality, the camcorder should have some provisions for connecting an external microphone. (You don't want your audience to think, "Gee, it'd be a great movie if it didn't have all that whirring and sneezing.") Most camcorders have a standard mini-jack connector for an external mic, and some high-end camcorders have a 3-pin XLR connector. XLR connectors — also sometimes called *balanced* audio connectors — are used by many high-quality microphones and PA (public address) systems.
- **Manual controls:** Virtually all modern camcorders offer automatic focus and exposure control, but sometimes (see Chapter 4) manual control is preferable. Control rings around the lens are easier to use than tiny knobs or slider switches on the side of the camera — and they'll be familiar if you already know how to use 35mm film cameras.

The spec sheet may try to draw your attention to various other camcorder features as well, but not all these features are as useful as the salesman might like you to believe. Features that seem exciting but are generally less important include

- **Night vision:** Some camcorders have an infrared mode that enables you to record video even in total darkness. Sony's NightShot is an example of this feature. If you want to shoot nature videos of nocturnal animals this may be appealing to you, but for day-to-day videography, it's less useful than you might think.
- **Still photos:** Many new digital camcorders can also take still photos. This is handy if you want to shoot both video and stills but don't want to lug along two cameras — but even relatively cheap digital still cameras take better photos than camcorders (even the most expensive ones).
- **USB port:** Some camcorders offer a USB connection in addition to FireWire. USB can be handy for transferring still photos into your computer, but I strongly recommend that you rely on FireWire for digital video capture. Many computer USB ports are not fast enough to handle full quality digital video.
- **Bluetooth:** This is a new wireless networking technology that allows various types of electronic components — including camcorders and computers — to connect to each other using radio waves instead of cables. Unfortunately the maximum data rate of current Bluetooth technology is still comparatively low (less than one megabit per second). In practical terms, that means Bluetooth won't be suitable for capturing digital video from your camcorder for the foreseeable future. A few camcorders incorporate Bluetooth technology anyway, and that may (or may not) come in handy if you still own the same camcorder a few years from now.
- **Built-in light:** If a camcorder's built-in light works as a flash for still photos, it at least serves a semi-useful purpose. But on-camera lights often have unfavorable lighting effects on your subjects; I recommend you rely on other light sources instead when you are shooting video.

And then there are some features which are essentially useless. Don't pay extra for these:

- **In-camera special effects:** Most digital camcorders boast some built-in effects. But why? Special effects can be added much more *effect*ively (so to speak) in your computer, using your editing software.
- **Digital zoom:** Digital zoom makes the image appear blocky and pixelated — again, why do it? I tend to ignore the big digital-zoom claims that camcorder manufacturers like to advertise. When you test the zoom feature on a camcorder, make sure you can *disable* digital zoom. Bottom line: You should be able to prevent the camera from automatically switching to digital zoom when you reach the optical zoom limit.

Accessorizing Your Camcorder

Few pieces of digital video gear are as underappreciated as camcorder accessories. Beyond the obvious things like a camcorder bag and spare tapes, there are a couple of extras which I feel are critical whenever you shoot video. These include

- **Extra batteries:** Your new camcorder should come with at least one battery, and I recommend buying at least one extra. Lithium Ion batteries are preferable to NiMH batteries because Lithium batteries last longer, and you don't have to worry about completely discharging them before recharging. (NiMH batteries have always been a pain because you were supposed to *completely* discharge them before recharging, and then you had to recharge them as fully as possible because NiMH batteries had a "memory" that would often prevent it from accepting a full charge.) A Lithium Ion battery, on the other hand, is more like the gas tank in your car: You can top it off whenever you want!
- **Lens cleaner:** Your camcorder's lens will inevitably need to be cleaned. Purchase a cleaning kit specifically recommended by your camcorder's manufacturer. This is important because the lens on your camcorder might have a special coating that can be damaged by the wrong kind of cleaner. Avoid touching the camcorder lens with *anything* (as much as humanly possible). I like to use canned air (available at computer supply stores) to blow dust or sand off the lens.
- **Lens hood:** Some high-end camcorders have hoods that extend out in front of the lens. A hood shades the lens surface to prevent light flares or other problems that occur when the sun or some other bright light source reflects directly on the lens. If your camcorder didn't come with a hood and your manufacturer doesn't offer one as an accessory, you can make a hood using black photographic paper tape (available at photographic supply stores). Make sure you check the camcorder's viewfinder, however, to ensure that your homemade hood doesn't show up in the video image!
- **Lens filters:** Filters fit onto the front of your lens and serve a variety of purposes. Some filters correct or modify the light that comes into the lens. *Polarizing filters* reduce reflections on glass or water that appear in a video shot. A *neutral density filter* improves color in bright sunlight. A *clear* or *UV (ultraviolet) filter* is often used simply to protect the camera's lens from getting dirty or scratched. Filters usually screw into a threaded fitting just in front of the lens, and you can purchase them from consumer electronics or photographic supply stores. If your camcorder doesn't accept a standard filter size (such as 37mm or 58mm), you will probably have to order filters specially designed for your camcorder by its manufacturer.

Some high-end cameras (such as the Canon GL2 or Sony DCR-VX2000) have built-in neutral density filters that you can turn on and off or adjust. This is a handy feature, but I still recommend that you always install at least a UV filter in front of your camcorder's lens to protect it from damage. A $15 UV filter is a lot easier and cheaper to replace than the glass lens in your camcorder. Tiffen (`www.tiffen.com`) sells a variety of camcorder filters; their Web site includes some excellent photographs that illustrate the effects of various lens filters on your video image.

Sounding Out Audio Equipment

All digital camcorders have built-in microphones, and most of them record audio adequately. You will probably notice, however, that the quality of the audio recorded with your camcorder's mic never *exceeds* "adequate." Most professional videographers emphasize the importance of good audio. They note that while audiences will tolerate some flaws in the video presentation, poor audio quality will immediately turn off your viewers. To record better audio, you have two basic options:

- Use a high-quality accessory microphone.
- Record audio using a separate recorder.

Choosing a microphone

If you want to connect a better microphone to your camcorder, the best place to start is with your camcorder's manufacturer — you'll need a *really* long cable. (Just kidding.) Usually accessory microphones are available from the manufacturer. These accessory units make use of connections, accessory shoes, and other features on your camcorder.

One type of special microphone you may want to use is a *lavalier* microphone — a tiny unit that usually clips to a subject's clothing to pick up his or her voice. You often see lavalier mics clipped to the lapels of TV newscasters. Some lavalier units are designed to fit inside clothing or costumes, though some practice and special shielding may be required to eliminate rubbing noises.

You might also consider a hand-held mic. These can be either held by or close to your subject, mounted to a boom (make your own out of a broom handle and duct tape!), or suspended over your subject. Suspending a microphone overhead prevents unwanted noise caused by breathing, rustling clothes, or simply bumping the microphone stand. Just make sure that whoever holds the microphone boom doesn't bump anyone in the head!

What's wrong with my camcorder's mic?

After spending hundreds if not thousands of dollars for a high-tech digital camcorder, you may be frustrated by the so-so quality of the audio recorded by the built-in microphone. The problem is that the camcorder's mic is prone to pick up a lot of noise you don't want, while not recording enough of the audio you actually want. Unwanted noise includes wind roar, people chatting next to or behind the camera, and even the camcorder's own tape drive. Consider the Chapter 3 sample clip on the CD-ROM that accompanies this book. In this clip, I shot a scene of my son watching sea lions at a local seaport. The camcorder's mic picked up the sound of barking sea lions, which is fine, but it also picked up voices of people chatting outside the video frame and various other undesirable sounds. If all I want in this shot is the sound of sea lions barking, I'll have to rely on another method of recording. Chapter 4 provides specific tips on recording better audio.

Microphones are generally defined by the directional pattern in which they pick up sound. The three basic categories are *cardioid* (which has a heart-shaped pattern), *omnidirectional* (which picks up sound from all directions), and *bidirectional* (which picks up sound from the sides). Figure 3-4 illustrates these patterns.

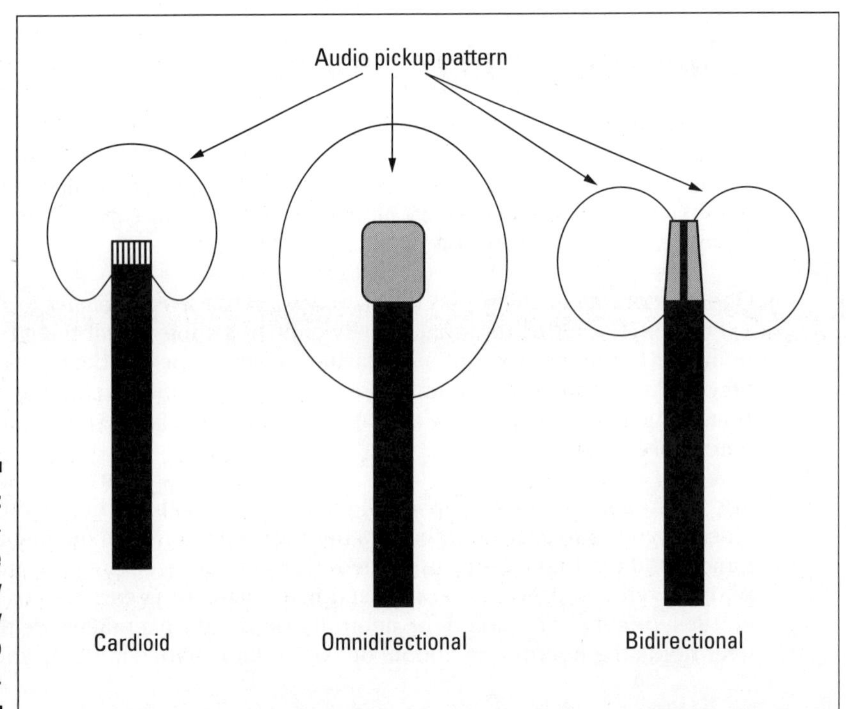

Figure 3-4: Microphones are defined by how they pick up sound.

A good place to look for high-quality microphones is at a musician's supply store. Just make sure that the connectors and frequency range are compatible with your camcorder or other recording device (check the camcorder's documentation). Finally, the Internet is always a good resource as well. One good resource is `www.shure.com`, the Web site of Shure Incorporated. Shure sells microphones and other audio products, and the Web site is an excellent resource for general information about choosing and using microphones.

Selecting an audio recorder

Separate sound recorders give you more flexibility, especially if you just want to record audio in a certain location but not video. Many professionals use DAT (digital audio tape) recorders to record audio, but DAT recorders typically cost hundreds or (more likely) thousands of dollars. Digital voice recorders are also available, but the amount of audio they can record is often limited by whatever storage is built in to the unit. For a good balance of quality and affordability, some of the newer MiniDisc recorders are good choices.

If you do record audio with a separate recorder, one problem you'll have later is precisely synchronizing the audio recording with the video image that you recorded. Professional video and filmmakers solve this problem using *slates*. A *slate* is that black-and-white board thingie with all the chalk writing on it that someone snaps closed just before the director yells, "Action!" The slate isn't just a kitschy Hollywood gimmick. When the slate is snapped closed in front of the camera, it makes a loud snapping noise picked up by all audio recorders on the set. That sound and the video image of the slate can be used later to precisely synchronize the separate audio and video recordings. If you plan to record audio with a separate audio recorder, I recommend that you construct and use a simple slate of your own. You can make it using two boards and a hinge purchased at any hardware store. (Just watch your fingers, okay?)

Shedding Some Light on the Scene

Most digital camcorders provide automatic aperture control (often called *exposure*). The *aperture* is the part of the camera that controls how much light is let in through the lens. It expands and contracts depending on light conditions, much like the iris in the human eye. But all the automatic controls in the world won't make up for a poorly lit scene. In Chapter 4, I provide some tips for properly lighting your scenes, and in Chapter 18, I recommend a couple of specific lighting products you may want to buy. But basically you are going to need several key bits of gear to better light your scene:

- **Lights:** Right about now, you're probably thinking, "No kidding." You can buy professional lights if you wish, but you don't need to spend hundreds of dollars to get good lights. Fluorescent shop lights are affordable and provide good-quality light, as are halogen work lights available at many hardware stores.
- **Backdrop material:** For some shots, you may want a backdrop behind your subject. You can make a backdrop frame out of pipe or cheap 1x3 pine boards (also known as *furring strips*) from your local lumber yard, and then tie, clamp, or staple the backdrop material to the frame. In Chapter 17, I show you how a backdrop using blue vinyl tablecloth material can be used to create a "bluescreen" special effect.
- **Clamps:** While you're at the hardware store buying lights and backdrop stuff, pick up some cheap spring clamps. Clamps can be used for holding backdrops together, holding lights in position, or playfully clipping unsuspecting crew members as they walk past.
- **Extension cords:** You'll need to plug in all your fancy lights somehow.
- **Duct tape:** If you can't do it with duct tape, it probably can't be done! I like to use duct tape to secure extension cords to the ground so that they aren't a trip hazard.
- **Translucent plastic sheets and cheesecloth:** Get these at your local art supply store to help diffuse and soften intense lights.
- **Reflective surfaces:** Use poster board, aluminum foil, or even plastic garbage bags to bounce light onto your subjects. Crumple foil to provide a more diffuse reflection, and tape the foil or plastic bags to boards so they're easier to handle.

Lights get hot, so use care when handling them after they've been in use for a while. Also, if you use plastic, cheesecloth, or other materials to diffuse light, position those materials so they aren't too close to hot lights.

Stabilizing Your Video Image

Although modern camcorders are small and easy to carry around, you'll probably find that most of your shots benefit from a tripod or other method of stabilization. Even the cheap $20 tripod that you got for free with your camcorder purchase is better than nothing for stationary shots. If you're looking for a higher-quality tripod, here are some features that can make it worth the extra cash:

- **Strong legs and bracing:** Dual-stanchion legs and strong bracing greatly improve the stability of the camera.
- **Lightweight:** The best tripod in the world doesn't do you any good if it's so heavy that you never take it with you. Better tripods use high-tech materials like aircraft aluminum, titanium, and carbon fiber to provide lightness without sacrificing strength.
- **Bubble levels:** Some tripods have bubble levels (like those carpenters use) to help ensure that the camera is level. Few things are more disorienting in a video shot than an image that is slightly skewed from level.
- **Counterweights:** Adjustable counterweights help you keep the tripod head and camera balanced even if you're using a heavy telephoto lens or other camcorder accessory.
- **Fluid heads:** This is probably the most important feature of a high-quality tripod. A fluid head enables you to pan the camera smoothly, eliminating the jerky panning motion associated with cheaper tripods.

Chapter 18 discusses some other stabilization devices that you may want to consider, especially if you shoot a lot of high-action video.

Part II
Gathering Footage

"I found these two in the multimedia lab, using iMovie to morph faculty members into farm animals."

In this part...

When you make your own movies, it's easy to get caught up in the process of editing — but before you can do that, you need something to edit. That means shooting, recording, or otherwise obtaining some source material. This part shows you how to shoot great video, record better audio, and import that audio and video into your computer.

Chapter 4

Shooting Better Video

In This Chapter

- Planning a video project
- Shooting better video
- Recording better audio
- Keeping a video-shoot checklist

One of the great things about digital video is how easy it is to edit. But all the editing in the world won't do you any good if you don't have decent source material from which to work. Shooting video is definitely an art form, but that doesn't mean you need to be naturally gifted or have years of media schooling to produce great movies. All you really need is some patience, a few helpful tips, and maybe a little creativity. Oh yeah, and tapes: Lots of blank videotapes. Not even the pros shoot perfect video every single time!

This chapter helps you shoot and record better audio and video. I'll show you how to plan a video project, and how to shoot video and record audio effectively. I've also included a handy checklist that you can use to make sure you've got all your gear when you're getting ready to go out for a video shoot.

Planning a Video Production

Camcorders are so simple to use these days that they encourage seat-of-the-pants videography, which isn't always the best idea. Just grabbing your camcorder and hastily shooting may be fine if you're shooting the UFO that happens to be flying overhead, but for most other situations, some careful planning will improve your movie.

The first thing you're probably going to do in any video project is shoot some video. Even if you are shooting a simple school play or family gathering, you can and should plan many aspects of the shoot:

- **Make a checklist of shots that you need for your project.** While you're at it, make an equipment checklist too. I've included a video-shoot checklist at the end of this chapter.
- **Survey the shooting location.** Make sure passersby won't trip over your cables or bump the camera. (It makes you unpopular, and can ruin your footage to boot.)
- **Talk to property owners or other responsible parties.** Identify potential disruptions, and make sure you have permission to shoot. For example, your kids' school probably doesn't mind if you shoot video of Suzie's band concert, but commercial concerts or sporting events usually have rules against recording performances.
- **Plan the time of the shoot.** This is especially important if you are shooting outside. What part of the sky will sunlight be coming from? Do you want to take advantage of the special light available at sunrise or sunset?
- **Bring more blank tapes and charged batteries than you think you'll need.** You never can tell what may go wrong, and preparing for the worst is always a good idea. When it comes to blank tapes and spare (charged) batteries, too many is always better than not quite enough.

Enlisting a crew

If you're like me, most of your video shoots will actually be pretty informal, so assigning a director, sound engineer, and key grip probably seems a little silly. Besides, people who are formally assigned to such jobs will want special T-shirts and name tags and lattes and all kinds of other stuff that isn't accounted for in your shoestring budget.

Still, you're probably going to need some help — with setting up equipment, holding lighting props and microphones, standing guard to prevent passersby from walking in front of the camera, and (of course) hauling gear. I've found that my kids are extremely helpful when it comes to videography because they *know* that moviemaking is cool. (Why are children always smarter than everybody else?) You'll want to take a few minutes to indoctrinate whomever you enlist for your video crew on the finer points of moviemaking and provide some dos and don'ts to follow while they're on the set. Helpers should be trained on any equipment they're going to use, and they should know where they can (and cannot) stand to avoid showing up in the picture. Anyone holding a microphone must be told that little hand movements may cause loud thumps and other noise; anyone holding a light reflector should be reminded to sit still lest strange shadows pan wildly back and forth on the scene.

Perhaps most importantly, remind all helpers that silence is golden. A camera lens may be limited to a specific field of vision, but a microphone isn't. When little Johnny comes over to you behind the camera and whispers, "Daddy, I have to go to the bathroom," his revelation will be recorded loud and clear on the tape.

If you want to shoot high-quality video, you'll be happy to learn that staging an elaborate video production with dozens of staff members, acres of expensive equipment, and professional catering is not necessary. But you can do some simple things to improve any shooting situation — whether you are casually recording a family gathering or producing your own low-budget sci-fi movie. The following sections help you shoot better video, no matter the situation.

Composing a Shot

Like a photograph, a great video image must be thoughtfully composed. Start by evaluating the type of shot you plan to take. Does the shot include people, landscapes, or some other subject? Consider what kind of tone or feel you want to achieve. Figure 4-1 illustrates how different compositions of the same shot can affect its overall tone. In the first shot, the camera looks down on the subject. Children are shot like this much too often: This approach makes them look smaller and inferior. The second shot is level with the subject and portrays him more favorably. The third shot looks up at the subject and makes him seem important, dominant, almost larger than life.

Figure 4-1: Composition can greatly affect how your subject is perceived.

Evaluating Lighting

For the purposes of shooting video, light can be subdivided into two categories: good light and bad light. Good light allows you to see your subject, and it flatters your subject by exposing details you want shown. Good light doesn't completely eliminate shadows, but the shadows don't dominate large portions of the subject either. Bad light, on the other hand, washes out color and creates *lens flares* (the reflections and bright spots that show up when the sun shines across the lens) and other undesired effects. Consider Figure 4-2: The right side of the subject's face is a featureless white glow because it's washed out by intense sunlight. Meanwhile, the left side of the face is obscured in shadow. Not good.

Figure 4-2: This is what happens when you don't pay attention to light.

How do you light your shots effectively? Remain ever aware of both the good light and the bad. If you don't have control over lighting in a location, try to compose the shot to best take advantage of the lighting that is available. The following sections provide more specific lighting tips.

Dressing your cast for video success

My guess is that most of your video "shoots" will actually be pretty informal affairs, where you basically record an event that was scheduled to happen whether you brought your camcorder or not. Thus, you may have a hard time convincing everyone who is attending that they should dress appropriately for video. But there definitely are some types of clothes that work better in video than others — and if you have any control at all over what the people in your video wear, try making these suggestions:

- **Avoid clothes with lots of thin parallel lines or stripes:** Thin parallel lines (like those you'd find on coarse corduroy or pinstripe suits) don't get along well with TV screens; they create a crawling or wavy visual effect called a *moiré pattern*.
- **Limit the use of very bright shades of red and blue.** Red is especially problematic because it tends to bleed into neighboring portions of the video image. This doesn't mean everyone in your movie should wear dark, drab colors, however. In the best of all possible shoots, your subjects' clothing is bright enough to lend some interest, but contrasts with the background somewhat so they don't get lost in the video image.

Choosing lights

Professional photographers and videographers typically use several different lights of varying type and intensity. Multiple light sources provide more control over shadows and image detail, and different kinds of lights have different affects. Lights that you'll use break down into three basic types:

- **Incandescent:** These are your good old-fashioned light bulbs like the ones Thomas Edison invented. Most of the light bulbs around your house are probably incandescent. Incandescent lights are usually cheap, but the light temperature is lower than with other types of lighting (meaning they are not as bright) and they usually provide wavering, inconsistent performance.
- **Halogen:** Okay, *technically* halogen lights are also incandescent, but they usually burn at a much higher temperature (and provide a more consistent light over their lifetimes) than do regular bulbs. Many professional video-lighting systems are tungsten-halogen lights, which means they have a tungsten filament passing through a sealed tube of halogen gas. Good, cheap halogen work lights also work well for video lighting — and they are generally available at tool and hardware stores. Halogen work lights also sometimes come with useful stands, though you can make your own stands using threaded plastic pipe and clamps from your friendly neighborhood hardware store.
- **Fluorescent:** Fluorescents also tend to give off a high-temperature light, though the temperature can vary greatly depending on the condition and age of the bulb. Fluorescent light is usually both very white and soft, making it ideal for video lighting. (In Chapter 17, for example, I use fluorescent lights to make sure that a bluescreen is brightly and evenly lit.) Fluorescent fixtures and bulbs can be purchased for less than $20 and can be easily suspended above your subject.

If you use fluorescent bulbs, let them warm up for a few minutes before shooting your video. This should prevent flicker. If the bulbs still flicker after they've warmed for a bit, try using new or different bulbs. Also, pay attention to how fluorescent lights affect your audio recordings. Fluorescent bulbs tend to produce a hum in audio recordings, so some practice and testing may be necessary. If fluorescent humming in your audio recording is a problem, record audio separately or use a different kind of light.

Bouncing light

Shining a light directly on a subject is not necessarily the best way to illuminate it. You can often get a more diffuse, flattering effect by bouncing light off a reflective surface, as shown in Figure 4-3. You can make reflectors out of a variety of materials, depending on what light effect you're looking for:

- **Poster board:** White poster board is a good, cheap material you can find just about anywhere. Thicker poster board is easier to work with because it's rigid (meaning it won't flop all over the place and make noise while a helper holds it). With some spray paint, paint one side of a poster board gold and the other silver. Experiment using each side to gain just the right light quality.
- **Aluminum foil:** Crumple a large sheet of foil, and then spread it out again and tape it to a backing board. Crumpling the foil provides a more diffuse reflection than you can get from flat foil, but its reflection is still highly effective.
- **Black plastic garbage bag:** Yep, black plastic bags are pretty good reflectors even though they have a dark color. As with foil, you can tape or staple the bag to a backing board to make it easier to work with.

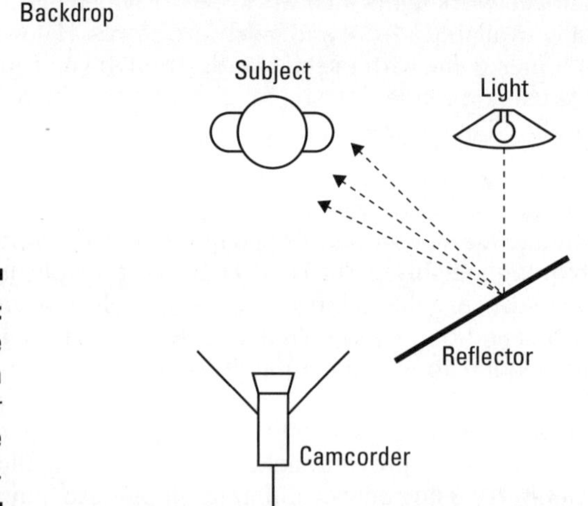

Figure 4-3: Bounce light off a reflector for a more diffuse light.

Reflectors can also be used to "fill" the lighting of your subject. Consider Figure 4-4: A light is directed at the subject to light up his face. This light is often called the *key light.* A reflector is positioned so some of the light that goes past the subject is bounced back onto the other side to fill in facial or other details.

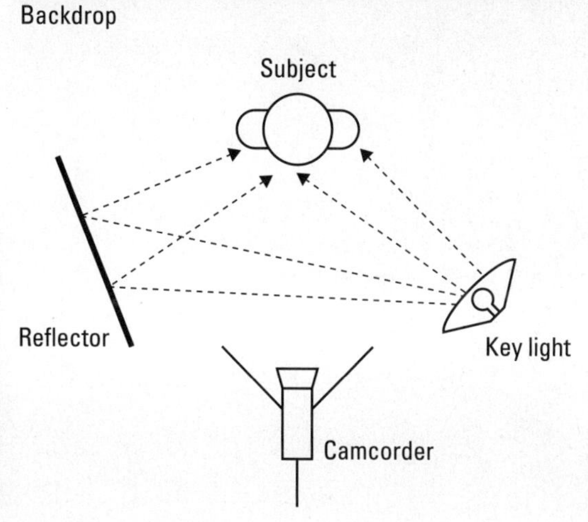

Figure 4-4: A reflector can be used to fill in lighting opposite the light source.

Diffusing light

Sometimes you may find that a light you're using to illuminate a subject is too intense. This is especially common with key lights such as the one shown in Figure 4-4. If all you're getting is a glaring white spot on your subject's face, you can diffuse the light by putting something between it and your subject:

- **Cheesecloth:** Available at art and cooking supply stores (and some grocery stores), cheesecloth has a coarse mesh and is useful both for diffusing light and straining beans or cottage cheese in the kitchen.
- **Translucent plastic:** Sheets of translucent plastic are also available at arts and craft stores. Professional videographers call these *gels.*

  Colored gels are often placed in front of lights that illuminate a backdrop for special lighting effects. If you do this, make sure you place a barrier between your key light and the backdrop (as shown in Figure 4-5) so the colored light from the gel isn't washed out by the white key light. A barrier may simply be a piece of cardboard that is held up by a stand or helper.

If you diffuse your light, you may have to move your lights closer to the subject. Experiment for the best results.

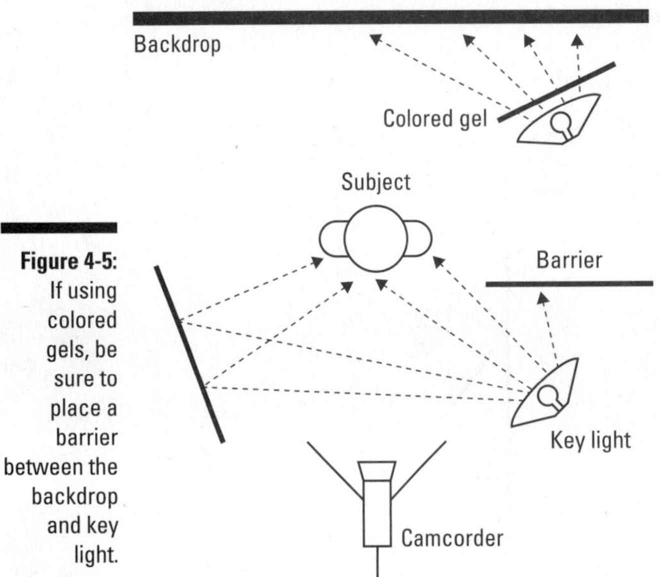

Figure 4-5: If using colored gels, be sure to place a barrier between the backdrop and key light.

Lights (especially halogen lights) tend to get very hot. To avoid fire hazards, you must use extreme care when placing gels or cheesecloth in front of lights. *Never* attach diffusers directly to lights. Position your diffusers some distance away from the lights so that they don't melt or catch on fire, and check the condition of your lights and diffusers regularly. Read and heed all safety warnings on your lights before using them.

Dealing with backlighting

Among the most common lighting problems you'll encounter when you shoot video are backlit situations. Backlit images occur when a relatively dark subject is shot against a relatively light background, as shown in Figure 4-6. The automatic exposure control in the camera adjusts exposure based primarily on that bright background, making the subject a dark, indistinguishable blob. To deal with a backlit situation

- **Avoid it:** When possible, try to shoot subjects so that they are not in front of a bright background, such as the sky.
- **Fight light with light:** Try to put more light on the subject. You may look silly toting a bright light around a sunny beach in the middle of the afternoon, but that is exactly what the pros do. If possible, try to shoot with the sun behind the camera.

- **Use camera settings:** Many camcorders have settings that automatically compensate for backlighting by increasing exposure. The results aren't always favorable, however, as you can see in Figure 4-7.

Figure 4-6: Who is that shadowy figure in this severely backlit situation?

Figure 4-7: Ah, there he is: now terribly overexposed by the camcorder's backlight compensator.

Your camera probably has settings to accommodate many special lighting situations. Always read the camcorder's documentation to see what features may be available to you — and practice using those features to see which ones work well and which ones don't.

Using lens filters

Your camcorder can probably accept some lens filters which screw on in front of the lens (see Chapter 3 for more on filters). Filters can be used to improve various lighting situations. For example, if you are shooting outdoors in a brightly sunlit location, you may find that colors look kind of washed out in the light. A *neutral-density filter* can compensate for the light and improve the way colors look.

Other lens filters can provide special lighting effects. For example, a *star filter* causes star patterns to shoot out from light points that appear in the image. This can give the scene a magical look. For more on how lens filters can change and improve the way your video images look, check out the Tiffen Web site at `www.tiffen.com`.

Controlling Focus and Exposure

Virtually all modern camcorders include automatic exposure and focus controls. Automation is really handy most of the time, but it's not perfect. If you always rely on auto focus, you will inevitably see the lens "hunting" for the right setting during some shots. This will happen a lot if you shoot moving subjects. Likewise, if you are shooting over a crowd or past other objects, the camera might focus on the closer objects instead of the desired subject. If your camera has a manual focus mode, you can avoid focus hunting by turning off auto focus.

Manual focus is pretty difficult to control if you're using a small dial or slider switch on the side of a camcorder. As I mention in Chapter 3, try to get a camera with a focus ring around the lens. This will make manual focus much easier to control.

I also urge you to learn how to use the manual exposure control (also called the *iris*). Exposure determines how much light is allowed to pass through the lens. It dilates and contracts much like the iris in a human eye. Manual exposure control allows you to fine tune exposure if the automatic control or camcorder presets aren't providing the desired light levels. Some higher end digital camcorders have a helpful feature called a *zebra* pattern. As you adjust exposure, a striped pattern will appear in overexposed portions of the image. Overexposed areas will appear as washed-out, colorless white blobs in your video image. A zebra pattern makes controlling exposure a lot easier: I have found that overexposing a video shot when you are manually adjusting exposure is very easy. For an example of an overexposed image, refer to Figure 4-7.

Although every camera is different, most camcorders have an infinite setting (∞) on the manual focus control. In most cases, anything that is more than about ten feet away will be in focus when the lens is set to infinite. Ten feet isn't a long distance, so you may be able to resolve many focus problems by simply using the infinite setting.

Setting Up Your Camcorder

Perhaps the most important tip I can give you before you shoot your video is this: Know your camera. Even today's least expensive digital camcorders are packed with features that were wildly advanced (and expensive) just a few years ago. Most digital camcorders include image stabilization, in-camera effects, and the ability to record 16-bit stereo audio. But these advanced features won't do you much good if they aren't turned on or configured properly. Spend a few hours reviewing the manual that came with your camcorder, and practice using every feature and setting. In particular, check the following:

- **Audio:** Many new camcorders are set by default to record only 12-bit audio, also sometimes called the 32-KHz (kilohertz) setting. Fire up your camcorder right now and make sure that it is set to 16-bit (48KHz) audio instead, and never change it back. 16-bit audio is higher quality, and it won't cause any problems later on when you want to capture video into your computer (or do pretty much anything else with it). For more on working with audio and understanding the bit and kilohertz settings, see Chapter 7. Audio recording is also described later in this chapter.
- **Focus and exposure:** In the previous section, I mention those times when you want to control focus and exposure manually. If you use manual focus or exposure control, switch them back to automatic before you turn off the camcorder. That way, the camcorder is ready for quick use later on when Bigfoot momentarily stumbles into your camp.
- **Special effects and exposure modes:** As with manual focus and exposure, if for some reason you use any of the built-in effects in your camera, make sure you disable them before turning off the camcorder so that it will be ready to go the next time you use it. The same thing goes for special exposure modes.

- **Stow the lens cap securely:** It seems obvious, I know, but if there is a clip or something that allows you to securely stow the camcorder's lens cap, use it. If you let the cap hang loose on its string, it will probably bang into the microphone and other parts periodically, making a lot of noise you don't want to record.
- **Use a new tape:** Even though digital video doesn't suffer from the same generational loss problems as analog video (where each play of the tape degrades the recording quality), various problems can still occur if you reuse digital tapes. Potential problems include timecode breaks (described later in this chapter) and physical troubles with the tape itself.

Keep the camcorder manual in your gear bag when you hit the road. It may provide you with an invaluable reference when you're shooting on location. Also, review the manual from time to time: Your camcorder no doubt has some useful or cool features that you forgot all about. If you've lost your manual, check the manufacturer's Web site. You might be able to download a replacement manual.

Shooting Video

Once your camcorder is configured and set up the way you want it, it's time to start shooting some video. Yay! One of the most important things you'll want to work on as you shoot video is to keep the image as stable as possible. Your camcorder probably has an image stabilization feature built in, but image stabilization can do only so much. I recommend using a tripod for all static shots, and a monopod or sling for moving shots. (See Chapter 18 for more on monopods and slings.)

Pay special attention to the camera's perspective. As I mentioned earlier (and demonstrated in Figure 4-1) the angle of the camera greatly affects the look and feel of the video you shoot. I often find that lowering the level of the camera greatly improves the image. Some high-end camcorders have handles on top that make shooting from a lower level easier. Virtually all digital camcorders have LCD panels that can be swiveled up so you can easily see what you're recording, even if you're holding the camera down low.

Be especially careful to avoid letting the camera roll to one side or the other. This skews the video image as shown in Figure 4-8, which is extremely disorienting to the viewer. Try to keep the camera level with the horizon at all times. The following sections give additional recommendations for shooting better video.

If you're shooting a person in a studio-like situation, complete with a backdrop and fancy lighting, provide a stool for your subject to sit on. A stool will help your subject remain both still and relaxed during a long shoot, and (unlike a chair) a stool will also help the subject maintain a more erect posture.

Figure 4-8: Don't let the video image get skewed like this — it's very disorienting.

Panning effectively

Moving the camera across a scene is called *panning*. You'll often see home videos that are shot while the person holding the camcorder pans the camera back-and-forth and up-and-down, either to follow a moving subject or to show a lot of things that don't fit in a single shot. This technique (if you can call it that) is called *firehosing* — usually not a good idea. Practice these rules when panning:

- **Pan only once per shot.**
- **Start panning slowly, gradually speed up, and slow down again before stopping.**
- **Slow down!** Panning too quickly — say, over a landscape — is a common mistake.
- **If you have a cheap tripod, you may find it difficult to pan smoothly.** Try lubricating the tripod's swivel head with WD-40 or silicon spray lubricant. If that doesn't work, limit tripod use to stationary shots. Ideally you should use a higher-quality tripod with a fluid head for smooth panning (see Chapter 18 for more on choosing a tripod).
- **If you're shooting a moving subject, try moving the camera with the subject, rather than panning across a scene.** Doing so reduces out-of-focus issues with the camera lens, and helps keep the subject in-frame.

Using (not abusing) the zoom lens

Most camcorders have a handy zoom feature. A zoom lens is basically a lens with an adjustable focal length. A longer lens — also called a *telephoto* lens — makes far-away subjects appear closer. A shorter lens — also called a *wide angle* lens — allows more of a scene to fit in the shot. Zoom lenses allow you to adjust between wide-angle and telephoto.

Because the zoom feature is easy to use and fun to play with, amateur videographers tend to zoom in and out a lot. I recommend that you avoid zooming during a shot as much as possible. Overuse of the zoom lens not only disorients the viewer, it also creates focal and light problems whether you're focusing the camera manually or using autofocus. Some zoom lens tips include

- **Avoid zooming whenever possible.** I know how tempting it is to zoom in on something cool or interesting in a video shot, but you should exercise restraint whenever possible.
- **If you must zoom while recording, zoom slowly.** You may need to practice a bit to get a feel for your camera's zoom control.
- **Consider repositioning the camera instead of using the zoom lens to compose the shot.** Wide-angle lenses (remember, when you zoom out you make the camcorder's lens more of a wide-angle lens) have greater *depth of field*. This means more of the shot is in focus if you're zoomed out. If you shoot subjects by zooming in on them from across a room, they may move in and out of focus. But if you move the camera in and zoom the lens out, focus will be less of a problem.

Avoiding timecode breaks

As I describe in Chapter 8, each frame of video is identified using a number called a *timecode*. When you edit video on your computer, timecode identifies the exact places where you make edits. On your camcorder, a timecode indicator tells you how much video has been recorded on the tape. This indicator usually shows up in the camcorder's viewfinder or the LCD panel. A typical timecode looks something like this:

```
00:07:18:07
```

This number stands for zero hours, seven minutes, eighteen seconds, and seven frames. If you have a 60-minute tape, timecode on that tape probably starts at 00:00:00:00 and ends at 00:59:59:29. In some cases, however, the timecode on a tape can become inconsistent. For example, suppose you record one minute of video, rewind the tape 20 seconds, and then start recording again. Depending on your camcorder, the timecode might count up to 00:00:40:00 and then start over at zero again. An inconsistency like this is called a *timecode break*. A timecode break is more likely to occur if you fast-forward a tape past a blank, unrecorded section and then start recording again.

When you capture video from a digital camcorder into your computer, the capture software reads the timecode from the tape in your camcorder. If the software encounters a timecode break, it will probably stop capture and be unable to capture any video past the break.

The best way to avoid timecode breaks is to make sure you don't *shuttle* the tape (fast-forward or rewind it) between recording segments. An alternative approach is to pre-timecode your tapes before shooting (as described in the sidebar "Blacking and coding your tapes" in this chapter). If you do have to rewind the tape — say, someone wants to see a playback of what you just recorded — make sure you cue the tape back to the end of the recorded video *before* you start recording again. Many camcorders have an *end-search* feature that automatically shuttles the tape to the end of the current timecode. Check your camcorder's documentation to see whether it has such a feature.

Blacking and coding your tapes

Years ago, computer geeks used lots of floppy disks for storing and moving files. Before they could be used, brand new disks had to be *formatted*, a simple process that allowed the computer to read and access the disk. Video pros also have a formatting process for new videotapes called *blacking and coding*. As the name suggests, blacking and coding is the process of recording black video and timecode onto new tapes. This helps ensure consistent timecode throughout the entire tape, thus avoiding potential timecode breaks.

You don't need special equipment to black and code new camcorder tapes. All you need is a lens cap. Place the lens cap on the camcorder and start recording from the start of the tape all the way to the end. You should do this in a quiet, darkened room in case the lens cap leaks a little light. As the black video is recorded onto the tape, consistent timecode is also recorded on the entire tape from start to finish. If consistent timecode is recorded on the entire tape, timecode breaks are not likely to occur when you use the tape to record real video later on.

Blacking and coding tapes is not absolutely mandatory with modern MiniDV camcorders: As long as you follow the basic recording guidelines mentioned earlier, you should be able to avoid timecode breaks. But if you do find that timecode breaks are a problem with your particular camera, blacking and coding all new tapes before you use them is a good idea. I especially recommend blacking and coding your tapes if you have a Sony Digital8 camcorder. Because Digital8 camcorders use Hi8 tapes, the camcorder may switch to analog Hi8 mode during playback or capture if you accidentally play or fast forward past the end of recorded digital timecode.

Recording Sound

Recording great-quality audio is no simple matter. Professional recording studios spend thousands (sometimes even millions) of dollars to set up acoustically superior sound rooms. I'm guessing you don't have that kind of budgetary firepower handy, but if you're recording your own sound, you can get pro-sounding results if you follow these basic tips:

- **Use an external microphone whenever possible.** The microphones built in to modern camcorders have improved greatly in recent years, but they still present problems. They often record undesired ambient sound near the camcorder (such as audience members at a play) or even mechanical sound from the camcorder's tape drive. If possible, connect an external microphone to the camcorder's mic input.
- **Eliminate unwanted noise sources.** If you *must* use the camcorder's built-in mic, be aware of your movements and other things that can cause loud, distracting noises on tape. Problem items can include a loose lens cap banging around, your finger rubbing against the mic, wind blowing across the mic, and the *swish-swish* of those nylon workout pants you wore this morning. I discuss ambient noise in greater detail in the following section.
- **Obtain and use a high quality microphone.** If you're recording narration or other audio in your "studio" (also known as your office) use the best microphone you can afford. A good mic isn't cheap, but it can make a huge difference in recording quality. The cheap little microphone that came with your computer probably provides very poor results.
- **Position the microphone for best quality.** If possible, suspend the mic above the subject. This way the microphone will be less likely to pick up noises made by the subject's clothes or bumping of the microphone stand.
- **Watch for trip hazards!** In your haste to record great sound, don't forget that your microphone cables can become a hazard on scene. Not only is this a safety hazard to anyone walking by, but if someone snags a cable, your equipment could be damaged as well. If necessary, bring along some duct tape to temporarily cover cables that run across the floor.

Earlier I mentioned that you should plan which video scenes you want to record. Planning the *audio* scenes that you want to record is also important. For example, suppose I want to make a video about a visit to the beach. In such a project I would like to have a consistent recording of waves crashing on the shore to use in the background. But if I record short, 5-to-10-second video clips, I'll never get a single, consistent *audio* clip. So, in addition to my various video clips, I plan to record a single, unbroken clip of the ocean.

Recording an unbroken clip of the ocean is exactly what I've done in the Chapter 4 sample clip on the CD-ROM that accompanies this book. The clip is about 45 seconds long and shows only the ocean. The video portion of the clip is a little boring (though if you've had a hectic day you may find it relaxing), but it will give me a 45-second audio clip of the ocean which will come in handy later.

Managing ambient noise

Ambient noise is the general noise that we don't usually think much about because it surrounds us constantly. Ambient noise might come from chirping birds, an airplane flying overhead, chattering bystanders, passing cars, a blowing furnace, the little fans spinning inside your computer, and even the tiny motor turning the tape reels in your camcorder or tape recorder. Although it's easy to tune out these noises when you're immersed in them, they'll turn up loud and ugly in your audio recordings later on.

If you're recording outdoors or in a public gathering place, you probably can't do much to eliminate the actual sources of ambient noise. But wherever you are recording, you can take some basic steps to manage ambient noise:

- **Use a microphone:** I know, this is about the millionth time I've said it, but a microphone placed close to your subject will go a long way towards ensuring that the sound you actually *want* to record is not totally overwhelmed by ambient noise.
- **Wear headphones:** Camcorders and tape recorders almost always have headphone jacks. If you plug headphones into the headphone jack you can listen to the audio that is actually being recorded, and possibly detect potential problems.
- **Shield the camcorder's mic from wind:** A gentle breeze may seem almost silent to your ear, but the camcorder's microphone may pick it up as a loud roar that overwhelms all other sound. If nothing else, you can position your hand to block wind from blowing directly across the screen on the front of your camcorder's mic.
- **Try to minimize sound reflection:** Audio waves reflect off any hard surface, which can cause echoing in a recording. Hanging blankets on walls and other hard surfaces will significantly reduce reflection.
- **Turn off fans, heaters, and air conditioners:** Air rushing through vents creates a surprising amount of unwanted ambient noise. If possible, temporarily turn off your furnace, air conditioner, or fans while you record your audio.

- **Turn off cell phones and pagers:** You know how annoying it is when someone's cell phone rings while you're trying to *watch* a movie; just imagine how bothersome it is when you're *making* a movie! Make sure that you and everyone else on the set turns those things off. Even the sound of a vibrating pager might be picked up by your microphones.
- **Shut down your computer:** Obviously this is impossible if you are recording using a microphone that is connected to your computer, but computers do tend to make a lot of noise, so shut them down if you can.
- **Warn everyone else to be quiet:** If anyone else is in the building or general area, ask them to be quiet while you are recording audio. Noises from the next room may be muffled, but they still contribute to ambient noise. Likewise, you may want to wait until your neighbor is done mowing his lawn before recording your audio.
- **Record and preview some audio:** Record a little bit of audio, and then play it back. This might help you identify ambient noise or other audio problems.

Creating your own sound effects

As you watch a TV show or movie, it is easy to forget that many of the subtle little sounds you hear are actually sound effects that were added in during editing, rather than "real" sounds that were recorded with the video image. This is often because the microphone was focused on the voice of a speaking subject as opposed to other actions in the scene. Subtle sounds like footsteps, a knock on the door, or splashing water are often recorded separately and added to the movie later. These sound effects are often called *Foley* sounds by movie pros, and someone who makes Foley sounds is called a *Foley artist*. Foley sound effects are named after audio pioneer Jack Foley, who invented the technique in the 1950s.

Recording your own sound effects is pretty easy. Many sounds will actually sound better in a video project if they're simulated, as opposed to recording the real thing. For example:

- **Breaking bone:** Snap carrots or celery in half. Fruit and vegetables can be used to produce many disgusting sounds, and they don't complain about being broken in half nearly as much as human actors.
- **Buzzing insect:** Wrap wax paper tightly around a comb, place your lips so that they are just barely touching the paper, and hum so that the wax paper makes a buzzing sound.

- **Fire:** Crumple cellophane or wax paper to simulate the sound of a crackling fire.
- **Footsteps:** Hold two shoes and tap the heels together followed by the toes. Experiment with different shoe types for different sounds. This may take some practice to get the timing of each footstep just right.
- **Gravel or snow:** Walk on cat litter to simulate the sound of walking through snow or gravel.
- **Horse hooves:** This is one of *the* classic sound effects. The clop-clop-clopping of horse hooves is often made by clapping two halves of a coconut shell together.
- **Kiss:** Pucker up and give your forearm a nice big smooch to make the sound of a kiss.
- **Punch:** Punch a raw piece of steak or a raw chicken.
- **Thunder:** Shake a large piece of sheet metal to simulate a thunderstorm.
- **Town bell:** To replicate the sound of a large bell ringing, hold the handle of a metal stew pot lid, and tap the edge with a spoon or other metal object. Experiment with various strikers and lids for just the right effect.

Some sound effects might be included with your editing software or are available for download. Pinnacle Studio, for example, comes with a diverse library of sound effects, which you can access by clicking the Show Sound Effects tab or choosing Album⇨Sound Effects. If the sound effects aren't currently listed, navigate to the folder `C:\Program Files\Pinnacle\Studio 8\Sound Effects`, and then open one of the thirteen sound-effect category folders such as `\Animals` or `\Squeaks`.

Apple iMovie 3 also includes some built-in sound effects. Open iMovie and click the Audio button. To view a list of sound effects, make sure you have iMovie Sound Effects listed in the menu at the top of the audio browser (as shown in Figure 4-9). You may also be able to download additional sound effects periodically from `www.apple.com/imovie/audio_effects.html`.

If you use Windows Movie Maker, you may be able to download free sound effects from Microsoft. Visit `www.microsoft.com/windowsxp/moviemaker/` and look for links to downloadable sound effects. If you are using another video-editing program, check the documentation or visit the publisher's Web site to see whether free sound effects are available.

Figure 4-9: Many video programs include sound effects you can use in your movies.

Video-Shoot Checklist for Dummies

A well-organized video shoot almost always results in better-quality video. But there are so many things to remember, you can easily forget a few things. Coming up are a couple of checklists to help you remember what you need to bring when you're preparing for a video shoot.

Basic checklist

This basic checklist includes important items that you'll need on virtually any shoot, no matter how informal:

- Camcorder
- Your camcorder owner's manual
- Extra *charged* batteries
- Spare *blank* tapes
- Lens cleaner
- Lens filters
- Tripod
- Duct tape
- Camcorder accessories (remote control, telephoto lens, and so on)
- Scene list

Advanced checklist

If you're planning a slightly more format video shoot, you'll probably want to bring the items on the basic checklist as well as these more advanced items:

- Lights
- Clothes/wardrobe for the cast
- Make-up and hair-care items
- Fans (to simulate wind and mess up that hair you just combed)
- Extension cords
- Reflectors
- Stool
- Backdrop
- Clamps
- AC adapter and/or battery charger for camcorder
- Microphone
- Audio recorder (don't forget batteries and blank audio tapes or MiniDiscs)
- Slate
- Script

Chapter 5

Capturing Digital Video

In This Chapter

- Getting ready to capture digital video
- Capturing digital video
- Troubleshooting video-capture problems

Software vendors all over the computer world have been rushing to offer programs that allow you to create and edit exciting movies on your computer. Video-editing programs have even become basic components of modern PC operating systems. Just as Microsoft Windows and the Apple Macintosh OS come with Internet browsers, e-mail programs, and even text editors, current versions of Windows and the Mac OS include free tools to help you make movies. Both Windows Me and Windows XP include a program called Windows Movie Maker, and Mac OS X comes with iMovie.

But all the video-editing software in the world doesn't do you much good if you don't have some actual video to edit. Thanks to FireWire (a new technology that allows computers and camcorders to easily share video) and a few other technologies, transferring video from the tape in your camcorder onto the hard disk in your computer is easy. This chapter shows you how. If you want to work with video that was shot with an analog camcorder — or you want to transfer video from a VHS videotape into your computer — check out Chapter 6.

Preparing for Digital Video Capture

The process of transferring video into your computer is often called *capturing*. Capturing digital video is pretty easy, but you should take some specific steps to ensure everything goes smoothly:

- **Install your hardware.** Your computer needs the right components to capture video — which means (among other things) having a FireWire or other capture card installed. (See Chapter 2 for more details on prepping your computer.)
- **Turn off unnecessary programs.** If you are like most people, you probably have several different programs running on your computer right now. Video capture requires a lot of available memory and processor power, and every running program on your computer uses some of those resources. E-mail, Web browser, and MP3 jukebox? Close 'em down. Cute desktop schemes and screen savers? Disable those too. I even recommend that you temporarily disable your antivirus software during video capture.

  If you're using Windows, take a look at the System Tray. (That's the area in the lower right corner of your screen, next to the clock.) Every little icon you see down there is a running program. Right-click each icon and close or disable as many of them as possible. Eventually (well, okay, ideally) your System Tray and Windows Taskbar looks something like Figure 5-1. You don't have to get rid of every single item, but do try to close or disable as many as possible. After all, it's a temporary arrangement. You should also disable your Internet connection during video capture as well. You can reactivate System Tray items later— including your antivirus software— by simply restarting your computer.

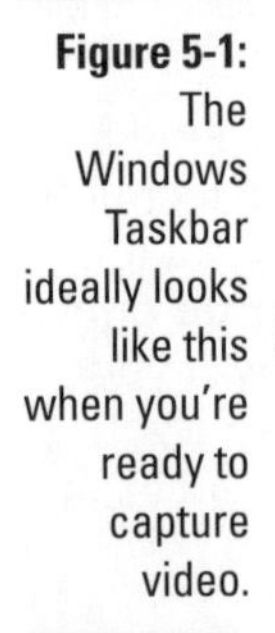

Figure 5-1: The Windows Taskbar ideally looks like this when you're ready to capture video.

  If you're using a Macintosh, look at the OS X Dock to make sure your programs are closed. (Any active program has a little arrow under it, as shown in Figure 5-2.) To quit a program, click its icon in the Dock and press ⌘+Q. The only icon you won't be able to quit is the Finder, of course.

Figure 5-2: Use the OS X Dock to make sure all open programs are closed.

- **Defragment your hard disk.** When your computer's operating system puts files on your hard disk, those files may wind up spread all over the place. This means that even if you have 60GB of free space, that 60GB might be broken up into little chunks here and there. This can cause trouble during video capture, especially with Windows machines, and *most* especially with version of Windows before Windows XP (such as Windows Me). Even if you have a Mac, it's still a good regular computer maintenance practice, and if you experience dropped frames during video capture, defragmentation can only help.

  I recommend you defragment your hard disk monthly, or right before video capture if the disk hasn't been defragmented recently. Defragmentation organizes the files on your hard disk so that the empty space will be in larger, more usable chunks. Some computer experts will probably tell you that defragmentation isn't as important with modern operating systems like OS X and Windows XP, but that advice does not really apply when you're working with video. Video is one of the few remaining tasks which still requires a defragmented hard disk.

  To defragment a hard disk in Windows, choose Start➪All Programs➪Accessories➪System Tools➪Disk Defragmenter. Choose the hard disk you want to defragment, and click Defragment (Windows XP) or OK (Windows Me and earlier).

  The Macintosh OS doesn't come with a built-in defragmenter, but you can use an aftermarket defragmentation utility such as Norton Disk-Doctor or Apple Disk First Aid.

- **Make sure you have enough hard-disk space.** I address this more in the next section, "Making room for video files."

Why is the process of transferring video to the hard disk called *capturing* instead of just *copying*? Even though the video is stored as digital data on your camcorder tape, that data must be turned into a computer file in order to be stored on your hard disk. *Capturing* is the process of turning video data into a computer file.

Making room for video files

Video files need a lot of space. Digital video typically uses 3.6 MB (megabytes) of space per second, which (if you do the math) means that 1GB (gigabyte) will hold about five minutes of video. You'll also need working space on the hard disk — so figure out approximately how much video you want to capture and multiply that number by four. For example, if you want to capture about 15 minutes of video, you'll need at least 3GB of disk space to store it. Multiply that by four (as a rule of thumb) to figure out that you should have at least 12GB of free space on your hard disk to get the job done.

When I say you need at least 12GB of space for 15 minutes of video, I do mean *at least*. Even though hard disk space may appear empty, your operating system — as well as your video-editing program — has to use that empty space periodically while working in the background. When it comes to hard-disk space, more really is better when you're working with video.

The first thing you need to do is figure out how much free space is available on your hard disk. In Windows, open My Computer (Start⇨My Computer). Right-click the icon representing your hard disk; choose Properties from the menu that appears. A Properties dialog box appears, similar to the one shown in Figure 5-3. It tells you (among other things) how much free space is available. Click OK to close the box.

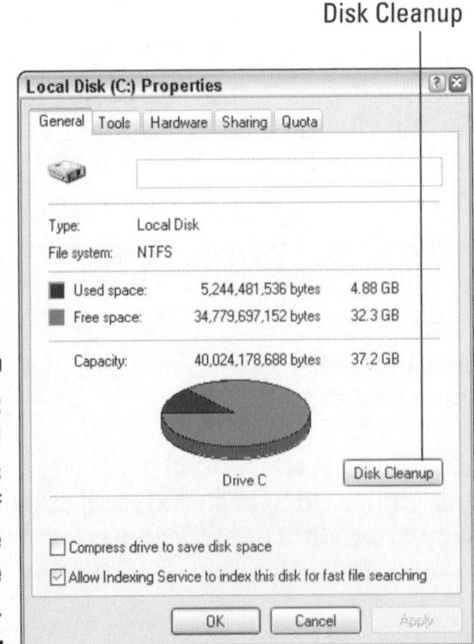

Figure 5-3: This hard disk has plenty of free space to capture some video.

On a Macintosh, click the icon for your hard disk just once; then press ⌘+I. An Info dialog box (similar to Figure 5-4) appears, showing you how much free space is available. Click the Close button or press ⌘+Q to close the Info dialog box.

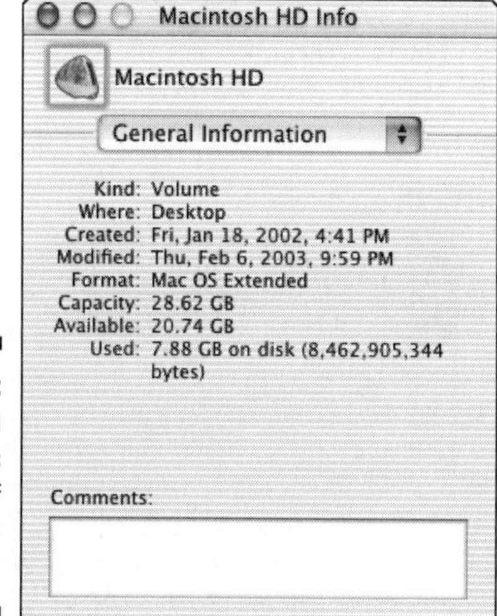

Figure 5-4: This hard disk has 20.74GB of free space available.

What if you need more space? Your options may be limited, but here are some things to consider:

- **Take out the garbage.** Empty the Recycle Bin (Windows) or Trash (Mac).
- **Clean up unneeded Internet files.** The cache for your Web browser could be taking up a lot of hard-disk space. The Windows Disk Cleanup utility (pictured in Figure 5-3) can help you get rid of these and other unnecessary files. On a Mac, you can empty the cache or control how much disk space is devoted to cache using the Preferences window for your Web browser.
- **Add a hard disk to your computer.** Adding a second hard disk to your computer can be a little complicated, but it's certainly one good way to gain more storage space. (See Chapter 2 for more on how to do this.)

Connecting a digital camcorder to your computer

Before you can edit video on your computer, you need to get the video *into* the computer somehow. Sorry, a shoehorn won't work — usually you connect a cable between your camcorder and the FireWire or USB port on your computer. Of the two, FireWire (also called IEEE-1394) is usually preferable.

FireWire ports have two basic styles: 6-pin and 4-pin. The FireWire port on your computer probably uses a 6-pin connector, and the port on your camcorder probably uses a 4-pin connector. This means you'll probably have to buy a 6-pin to 4-pin FireWire cable if you don't already have one. Figure 5-5 illustrates the differences between FireWire connectors.

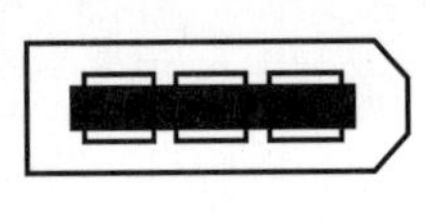

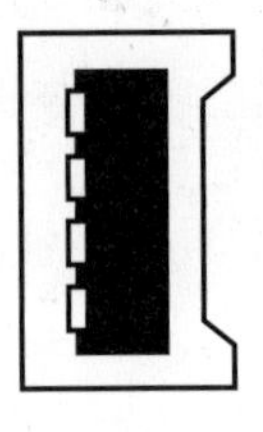

6-pin 4-pin

Figure 5-5: The two styles of FireWire connectors are 6-pin and 4-pin.

The FireWire connector on your camcorder may not be labeled "FireWire." It may be called IEEE-1394 (the "official" term for FireWire), DV, or i.Link. Whatever it's called, if the camcorder is digital and it has a port that looks like one of the connectors in Figure 5-5, it's FireWire-compatible.

FireWire is a "hot-swappable" port technology. This means you don't have to turn the power off on your computer when you want to connect a camcorder or other FireWire device to a FireWire port. Just plug it in, turn it on, and away you go. (Meanwhile, you may want to practice saying buzzwords like "hot-swappable" with a straight face.)

Connect the FireWire cable between the camcorder and your computer, and then turn on the power on your camcorder. Your camcorder probably has two power modes. One is a *camera* mode, which is the mode you use when you shoot video. The second mode is a *player* or *VTR* mode. This second mode is the one you want to turn on when you prepare to capture video from the camcorder's tape into your computer.

When you turn on the camcorder, your Windows PC will probably chime and then display the Digital Video Device window, which asks you what you want to do. This window is one of those handy yet slightly-annoying features that tries to make everything more automatic in Windows. Click Cancel to close that window for now.

One really cool feature of FireWire is a technology called *device control.* In effect, it allows software on your computer to control devices that are connected to your FireWire ports — including digital camcorders. This means that when you want to play, rewind, or pause video playback on the camcorder, you can usually do it using Play, Rewind, Pause, and other control buttons in your Pinnacle Studio or Apple iMovie application.

Some camcorders can use USB (*Universal Serial Bus*) ports instead of FireWire ports. The nice thing about USB ports is that — unlike FireWire ports — virtually all PCs made in the last five years have them. The not-so-nice thing is that most USB ports are too slow to handle full-quality video. Pinnacle Studio and Apple iMovie can capture from USB cameras using the same procedures described in this chapter for FireWire capture, but you may find that the video quality is reduced. Thus, I recommend that you use FireWire whenever possible.

Capturing Digital Video

Modern video-editing software makes capturing really easy. In the following sections, I show you how to capture video using Pinnacle Studio (on a Windows PC) and iMovie (on a Mac). Fortunately, the capture process in most programs is pretty similar, so you should be able to follow along no matter what software you are using.

Even if your FireWire card came with a separate capture utility, I recommend that you capture video in the program you plan to use for editing whenever possible. That way you won't have to go through the trouble of figuring out how to get the video from that capture utility into your desired editing program.

Capturing video in Pinnacle Studio

Pinnacle Studio for Windows offers an amazing level of moviemaking power for its price. The software is available by itself for a suggested retail price of $99, or you can buy it bundled with digital (FireWire) or analog video-capture cards for about $30 more. The top-of-the-line Studio package is Pinnacle Studio Deluxe (which retails for about $299), which includes the Studio software, capture hardware for both digital and analog video, and the Hollywood FX plug-in (which adds advanced video-editing tools to the basic Studio package).

A trial version of Studio can be installed from the companion CD-ROM for this book. See Appendix A for details.

To begin the capture process in Studio, click the Capture mode button or choose View⇨Capture. The Capture mode appears, as shown in Figure 5-6.

Setting capture options

Pinnacle Studio has some really cool options for capturing video. Unlike most affordable video-editing software, Studio supports online and offline editing. Offline editing is an advanced editing style that I described in Chapter 1. Studio facilitates offline editing using the Preview-quality capture mode. When capturing video with Studio, you can choose one of three basic quality settings:

- **DV full-quality capture:** Choose this option if you plan to export your movie back to videotape — and you have a generous amount of hard-disk space to use.
- **MPEG full-quality capture:** Choose this option if you plan to output your video to a VCD, S-VCD, DVD, or the Internet. MPEG capture can be customized further using some sub-options described in the next section.

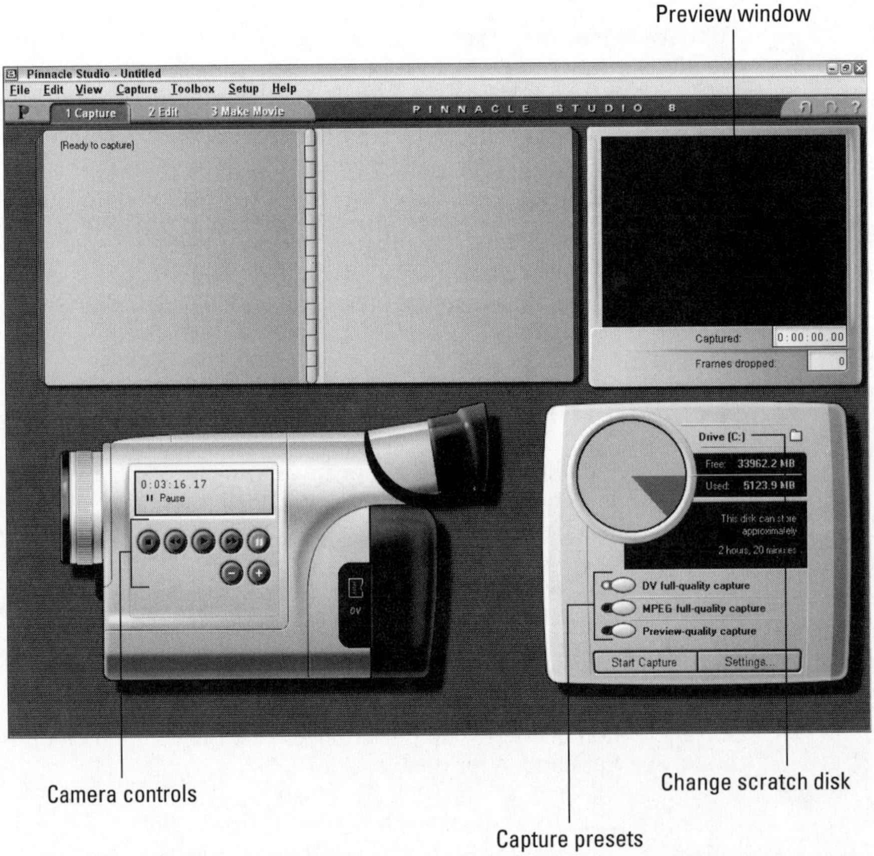

Preview window

Camera controls

Change scratch disk

Capture presets

Figure 5-6: The Studio Capture mode provides a friendly interface for capturing video.

- **Preview-quality capture:** Choose this option if you want to capture a lot of video to your hard disk but storage space is a concern. With this option, Studio captures video at a lower quality, which results in a smaller file size. Later, when you're done editing, Studio recaptures — at full DV quality — only the portions you want for your final movie; all your edits are automatically applied to the full-quality video.

If you're not quite satisfied with the quality of the preview you get, you can customize Preview-quality capture with sub-options described later in this chapter.

In addition to the three basic quality settings, Studio provides you with a variety of other capture options. You can find them in the lower-right portion of the screen (as shown in Figure 5-6). One of the things you can customize is the storage location for your captured video, often called the *scratch disk.* Click the folder icon to review the scratch disk folder and change it if you wish. For example, you may have a second hard disk that you wish to use for video storage. The default location for your scratch disk is

```
My Documents\Pinnacle Studio\Captured Video
```

This folder is created automatically when you install Studio on your computer.

You can also specify a variety of other capture settings. Click the Settings button in the lower-right corner of the screen to open the Pinnacle Studio Setup Options dialog box. Click the Capture Source tab to bring it to the front, as shown in Figure 5-7.

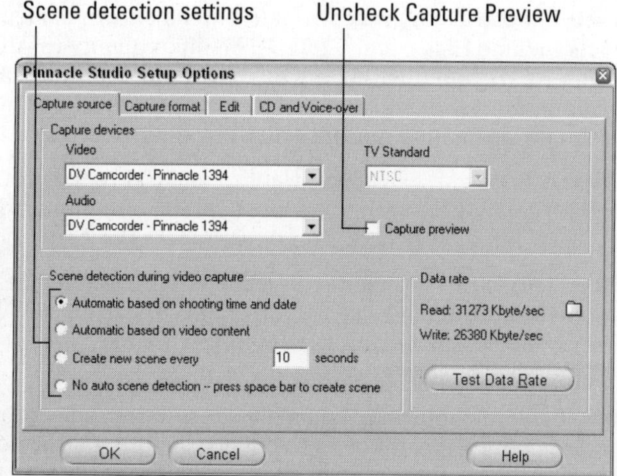

Figure 5-7: Choose basic capture options here.

This dialog box contains a lot of options and settings, but only two of them are really important for right now. First of all, I strongly recommend that you remove the check mark next to Capture Preview. When this option is enabled, Studio shows a preview of the video on-screen as you capture it. This preview uses up valuable memory and processor power that is better devoted to the actual video-capture process. You will still be able to view the video on the LCD display or viewfinder of your camcorder, so the on-screen preview is redundant anyway.

Second, review the scene detection settings. Studio can automatically detect when one scene ends and another begins and automatically turn each scene into a separate video clip. This comes in handy during editing, but if you don't like the feature, you can click the No Auto Scene Detection — Press Space Bar to Create Scene radio button. Otherwise, I recommend you keep it on the default setting as shown in Figure 5-7.

After you have changed these settings, click the Capture Format tab to bring it to the front. If you are using the DV capture preset, you won't be able to customize any of the settings on this tab. But if you are using the MPEG or Preview preset, you will be able to adjust some options as described in the next two sections.

Checking MPEG capture settings

If you're using the MPEG preset (your current preset is shown in the drop-down list in the upper left corner of the Capture Format tab), you can choose a standard group of settings or customize your settings (as shown in Figure 5-8). MPEG capture settings that you can adjust include these:

- **Sub-Preset:** Choose a sub-preset from the menu shown in Figure 5-8. Sub-presets include High quality (DVD), Medium quality (SVCD), Low quality (Video CD), and Custom. If you choose Custom, you can modify the remaining settings. If you choose the High, Medium, or Low quality sub-presets, the remaining options will by grayed out.
- **MPEG Type:** Choose MPEG1 if you want to make sure that your final movie will be compatible with the widest variety of computers, or choose MPEG2 for slightly better quality.
- **Resolution:** This is the screen size in width and height in pixels of the image you will capture. The size for full-quality DV is 720 x 480. Smaller sizes mean a smaller image, but it also means the files sizes for your video will be much smaller.
- **Pre-Filter:** If you're capturing at a smaller resolution, select this check box to improve the appearance of the image slightly (at the cost of absolute sharpness).
- **Fast Encode:** This option speeds up the capture process, but it can reduce quality.

- **Data Rate:** You can fine-tune quality and file size by adjusting data rate. To adjust data rate, move the slider back and forth. Lower data rates mean smaller files but also lower quality. In most cases, I recommend you keep the default data rate setting.
- **Include Audio:** If you only want to capture the video image from your tape and not the audio, uncheck this option.
- **MPEG Capture:** This menu helps you tailor capture to the speed of your computer. The safest option is to simply leave Use Default Encoding Mode selected. If your computer is very fast (2 GHz or faster processor), you can make capture more efficient with the Encode in Real Time option. If your computer is slower (slower than 1 GHz processor), choose Encode After Capturing if you encounter dropped frames or other problems during capture.

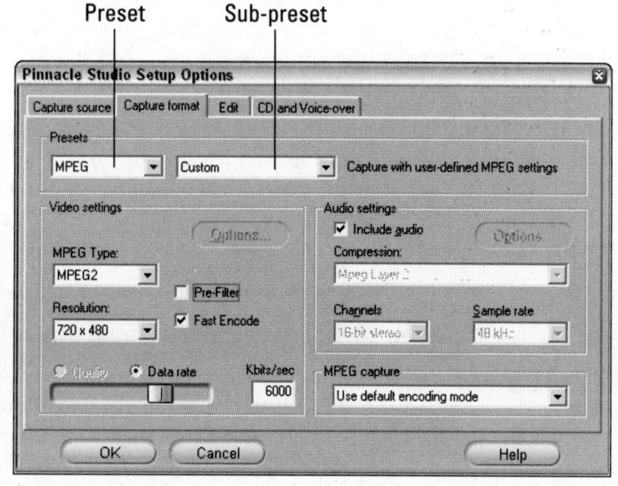

Figure 5-8: Choose Capture format options here.

When you're done setting capture format options, click OK to close the Pinnacle Studio Setup Options dialog box.

Previewing your capture settings

Pinnacle Studio's Preview capture mode is an excellent tool because it lets you store more source material on your hard disk for editing purposes without using up quite so much disk space. Like MPEG capture, if you choose the Preview preset in the main capture window (Figure 5-6) and then click Settings, you'll have a group of sub-presets to choose from on the Capture Format tab (Figure 5-8). It's usually best to just use one of the sub-presets, but if you choose Custom quality, you can adjust the following settings:

- **List All Codecs:** Choose this option to list all of the codecs that are installed on your system. (*Codecs* are compressor/decompressor schemes for audio and video, described in greater detail in Chapter 14.) I really don't recommend using this option because some of the codecs installed on your computer may not be compatible with the Pinnacle Studio software.
- **Compression:** Here you can choose a specific codec if you wish. For most preview captures, the Intel Indeo Video R3.2 or PCLEPIM1 32-bit Compressor codecs are fine.
- **Width and Height:** Choose a custom size for the video if you wish. Remember, the smaller the size, the less hard-disk space will be needed. If you're using a PIM1 codec, I recommend a frame size of 352 x 240 or smaller. For Indeo codecs, use 360 x 240 or smaller.
- **Frame rate:** The default frame rate for NTSC video is 29.97, but your files will be much smaller if you choose 14.985. If you're working with PAL video, you can choose a frame rate of 25 or 12.5. A lower frame rate means the video image won't be quite as smooth, but because it's only preview quality, this usually isn't a big deal.
- **Quality or Data Rate:** Select either the Quality or Data Rate radio button, and then use this slider to adjust the quality or data rate for the capture. It doesn't really matter if you choose Quality or Data Rate; the end result will be the same. Remember, lower quality (or data rate) means smaller file sizes.
- **Include Audio:** Deselect this option if you only want to capture video.
- **Channels:** Choose between 16-bit stereo (better audio quality) or 16-bit mono (smaller files).
- **Sample Rate:** You can probably say it with me by now: Higher sample rates provide better quality, lower sample rates mean smaller files.

When you're done setting capture format options, click OK to close the Pinnacle Studio Setup Options dialog box.

Capturing video

When you've finally got your capture settings just the way you want them, you're ready to capture. To do so, simply follow these steps:

1. **Connect your camcorder to your FireWire port as described in the previous section.**
2. **In Pinnacle Studio, click the Capture tab near the top of the window, or choose View⇨Capture.**
3. **Configure your capture options (as described earlier in this chapter).**

4. **Use the camera controls (as shown in Figure 5-6) to shuttle the camcorder tape to the beginning of the spot where you want to start capturing video.**

   Shuttle is another fancy term that video pros like to use when they talk about moving a videotape. When you rewind or fast forward a tape, you are shuttling it. See: You're already a video pro and you didn't even know it!

5. **Click the Start Capture button.**

   The Capture Video dialog box appears, as shown in Figure 5-9.

6. **Enter a name for the capture; this name will be used as the filename for the captured video later. If you want, enter a time limit for the capture.**

   By default, the time shown reflects the amount of free space on your hard disk. In Figure 5-9, you can see that my hard disk has enough room to store 607 minutes and 21 seconds of video using my current quality settings. It's usually safe to just leave this number alone unless you want Pinnacle to automatically stop capturing after a certain amount of time. For example, if I know that I only want to capture the first five minutes, I can enter 5 in the minutes field and 0 in the seconds field. Then I can go and get a cup of coffee or do something else without having to hurry back to manually stop capturing at some point. Pinnacle will automatically stop capturing after five minutes have gone by.

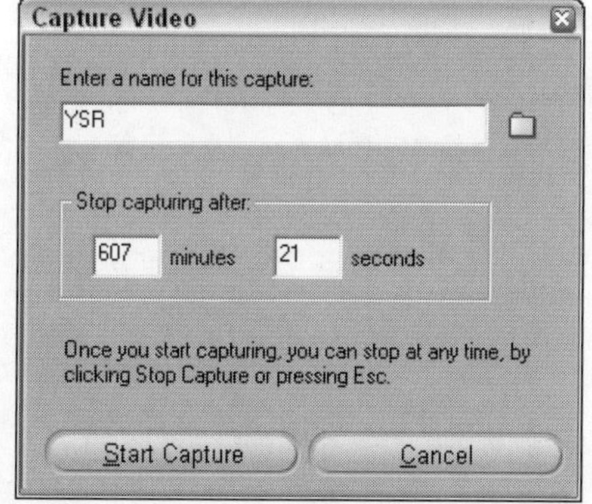

Figure 5-9: Enter a name and capture limit here.

7. **Click Start Capture.**

   Studio automatically starts playing your camcorder and capturing video.

8. **When you want to stop capturing, click the Stop Capture button or press Esc on your keyboard.**

As Studio captures your video, keep an eye on the Preview window, even if you have disabled on-screen preview in the capture settings. The Frames Dropped field should remain at zero. Dropped frames are a serious quality problem, but they can often be resolved. If you drop some frames, see the section entitled "Frames Drop Out During Capture" later in this chapter.

Capturing Video in Apple iMovie

Apple's iMovie doesn't offer quite as many capture options as Pinnacle Studio, but the capture process is simple and effective nonetheless. In fact, you can set just two capture options. To adjust capture preferences, choose iMovie⇨Preferences. The Preferences dialog box appears as shown in Figure 5-10. The two options relating to video capture are as follows:

- **New Clips Go To:** This default setting sends incoming clips to the Clips Pane. This is the best place to send new clips unless you want to quickly convert your imported video into a movie without any editing. What's the fun in that?
- **Automatically Start New Clip at Scene Break:** iMovie automatically recognizes when one scene ends and a new one begins. This useful feature often makes editing easier, so I recommend that you leave this option checked.

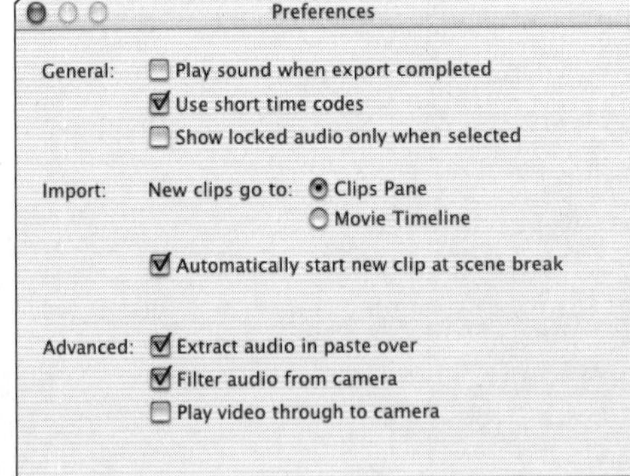

Figure 5-10: Import options also apply to captured video.

The iMovie interface is simple and easy to use. To capture video, follow these steps:

1. **Connect your camcorder to the FireWire port as described earlier in this chapter.**
2. **Switch the camera to VTR mode.**

3. **In iMovie, click the Camera button to switch iMovie to the Camera mode.**

   The Preview pane displays the message `Camera Connected` (as shown in Figure 5-11).

4. **Use the camera controls to identify a portion of video that you want to capture.**

   When you are ready, rewind the tape about ten seconds before the point at which you want to start capturing.

5. **Click the Play button in the camera controls.**
6. **Click the Import button when you want to start importing.**
7. **Click the Import button again when you want to stop importing.**

It's just that simple. Your captured clips automatically appear in the Clips Pane, where you can then use them in your movie projects.

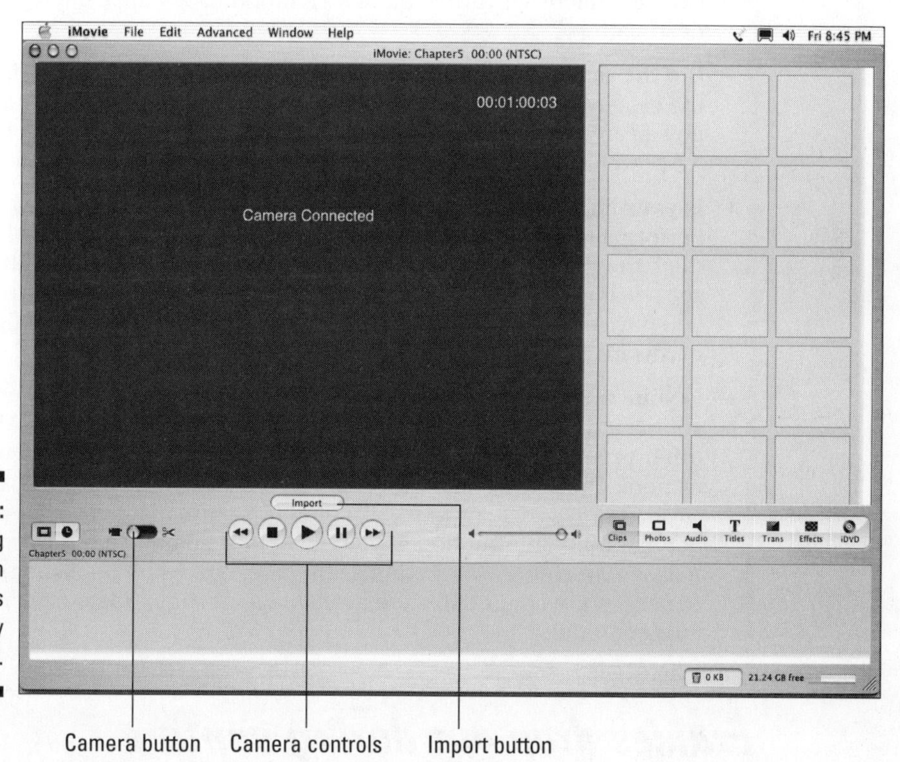

Figure 5-11: Capturing video in iMovie is pretty simple.

Troubleshooting Capture Problems

Video capture usually works pretty easily and efficiently with modern hardware, but some problems can still occur. Here's a quick tour of some common digital-capture problems — and their potential solutions.

You can't control your camera through your capture software

When you click Play or Rewind on the camera controls in your video-capture software, your digital camcorder *should* respond. If not, check the following items:

- **Check all the obvious things first:** Are the cables connected properly? Is your camcorder turned on to VTR mode? Does the camera have a dead battery?
- **Did the camera automatically power down due to inactivity?** If so, check the camera's documentation to see if you can temporarily disable the power saver mode. Also, consider plugging the camera in to a charger or AC power adaptor so that you aren't just running on battery juice.
- **Is your FireWire card installed correctly?** Open the System icon in the Performance and Maintenance section of the Windows Control Panel. Click the Hardware tab, and then click the Device Manager button. If you see a yellow exclamation mark under IEEE 1394 Bus host controllers, you have a hardware problem. See Chapter 2 for more on configuring hardware.
- **Is your camcorder supported?** Most modern digital camcorders are supported by Apple iMovie, Pinnacle Studio, Windows Movie Maker, and other programs. But if the software just doesn't seem to recognize the camera, check the software vendor's Web site (`www.apple.com`, `www.pinnaclesys.com`, or `www.microsoft.com`, respectively) for camera compatibility information. If your camera is so new that it wasn't originally supported by your editing software, check the publisher's Web site to see if software updates are available to accommodate newer camcorder models.

Frames drop out during capture

Video usually has about 30 frames per second, but if the capture process doesn't go smoothly, some of those frames could get missed or *dropped,* as video pros call it. *Dropped frames* show up as jerky video and cause all kinds of other editing problems, and usually point an accusing finger at your hardware, for one of four possible reasons:

- **Your computer isn't fast enough:** Does your computer meet the system requirements that I recommend in Chapter 2? How fast is the processor? Does it have enough RAM?
- **Your computer isn't operating efficiently:** Make sure all unnecessary programs are closed as I described earlier in this chapter. You may also be able to tweak your computer's settings for better video-capture performance. Check out the OS Tweaks in the Tech Support section of `www.videoguys.com` for tips and tricks for helping your computer make the best possible use of its resources.
- **You left programs running:** Make sure that your e-mail program, Internet browser, music jukebox, and other programs are closed. I also recommend that you disable your Internet connection during video capture, even if you have a broadband (cable modem, DSL) connection. When your connection is active, you probably have a few utilities running in the background looking for software updates to your operating system and other programs. A well-meaning message that updates are available for download might appear right in the middle of video capture, causing dropped frames.
- **Your hard disk can't handle the data rate:** First off, does your hard disk meet the requirements I outline in Chapter 2? If so, maybe some hard-disk maintenance is in order. First of all, the more stuff that is packed onto your hard disk, the slower it will be. This is one reason I recommend (pretty often, in fact) that you *multiply the amount of space you think you'll need by four.*

To keep your hard disk working efficiently and quickly, defragment your hard disk periodically (at least once a month). In Windows, choose Start➪All Programs➪Accessories➪System Tools➪Disk Defragmenter. On a Mac, you'll need to obtain a third-party disk-maintenance tool such as Norton Disk Doctor.

If you're using Pinnacle Studio, you can use that software to test your hard disk's data rate. In Studio, choose Setup➪Capture Source. In the Setup Options dialog box that appears, click the Test Data Rate button. Studio tests your data rate and lists the read, write, and Max safe speeds. The Max safe speed is a speed which Studio determines is necessary for glitch-free video capture. For DV-format video, the Max safe speed should be at least 4000KB per second.

Capture stops unexpectedly

If the capture process stops before you want it to, your culprit could be mechanical. Check the following:

- **Did you forget to rewind the camcorder tape?** This is a classic "oops" that happens to nearly everybody sooner or later.
- **Is your hard disk full?** This is bad juju, by the way. Try to avoid filling up your hard disk at all costs.
- **Is there a timecode break on the tape?** Inconsistent timecode on a digital videotape can create all sorts of havoc when software tries to capture video. (See Chapter 4 for tips on avoiding timecode breaks.)
- **Did Fluffy or Junior step on the Esc key?** It happens. My cat has fouled more than one capture process. (Hey, biomechanical still counts as "mechanical," right?)

Chapter 6

Capturing Analog Video

In This Chapter

- Setting up your computer for capture
- Getting the right kind of capture hardware
- Optimizing your PC's settings for capture
- Capturing analog video

After all this talk about how great digital video is, you may be wondering why I devote an entire chapter to analog video. Most people assume that digital video is the new hotness, and that analog video is just old and stinky.

Analog video *is* old and stinky, but you might still have some really good old video footage on analog tapes. Your daughter in middle school isn't going to be taking those first baby steps again, so if you have it on analog, you probably want a way to convert it to digital. If you have or used to have a VHS, VHS-C, S-VHS, 8MM, or Hi-8 camcorder, you can still capture video shot with those cameras into your computer, where you can edit it, put it online, or even burn it on a DVD. All you really need is the right hardware. This chapter shows you how to use ye olde analog video with your brand new (or even not-so-new) computer.

Preparing for Analog Video Capture

In Chapter 5, I showed you how to capture digital video into your computer. If you have the right hardware — a digital camcorder and a FireWire port — digital video is really easy to work with. Analog video can also be easy to capture *if you have the right hardware*. The next few sections help you prepare to capture analog video into your computer.

Unlike digital video, analog video is not compatible with computers. This is because all of the data that computers work with is digital. Before you can use analog video in your computer, it must be converted to digital, or *digitized*. Analog video-capture hardware digitizes the video as it is captured.

Preparing your computer for analog video

Getting your computer ready to capture analog video is a lot like getting ready to capture digital video. Before you can capture analog video, you have to

- **Set up your capture hardware.** I'll show you what hardware you need and how to set it up in the following section, "Setting up capture hardware."
- **Turn off unnecessary programs.** Whether you're working with analog or digital video, your computer will work more efficiently if you close all programs that are not needed for the actual capture process. See Chapter 5 for more on making sure all unnecessary programs are closed on your Macintosh or Windows PC.
- **Make sure there's enough free space on your hard disk.** Five minutes of digital video uses about 1GB (gigabyte) of hard-disk space. Unfortunately, there is no single, simple formula for figuring out how much space your analog video will require.

  Digital video recorded with a MiniDV, MicroMV, or Digital8 camcorder uses the DV codec, which uses a steady 200MB (megabytes) per minute of video. A *codec* — short for compressor/decompressor — is the software scheme used to compress video so it fits reasonably on your computer. (Codecs are described in greater detail in Chapter 13.) Some analog-capture devices let you choose from a list of different codecs to use during video capture; many codecs have settings you can adjust. Which codec you use (and the settings you select) can greatly affect both the quality of your capture video and the amount of space it uses up on your hard disk. I show you how to adjust capture settings later in this chapter in the section, "Adjusting video capture settings."

  Fortunately, most analog video-capture programs make it pretty easy to determine whether you have enough hard disk space. The Pinnacle Studio capture window, for example (shown in Figure 6-1), shows you exactly how much free space is available on your hard disk, and it gives you an estimate of how much video you can capture using the current settings.

  In Figure 6-1, you can see that the Pinnacle software is telling me that I can squeeze about two hours and ten minutes of video onto my hard disk. But (as I mention in Chapter 5) you need to leave some free space on your hard drive to ensure that your video-editing software and operating system can still work efficiently.

Setting up capture hardware

Before you can capture analog video, you need to have some way to connect your VCR or analog camcorder to your computer. To do so, you have two basic options. You can either use an analog-capture card or an external video converter. I describe each option in the next two sections.

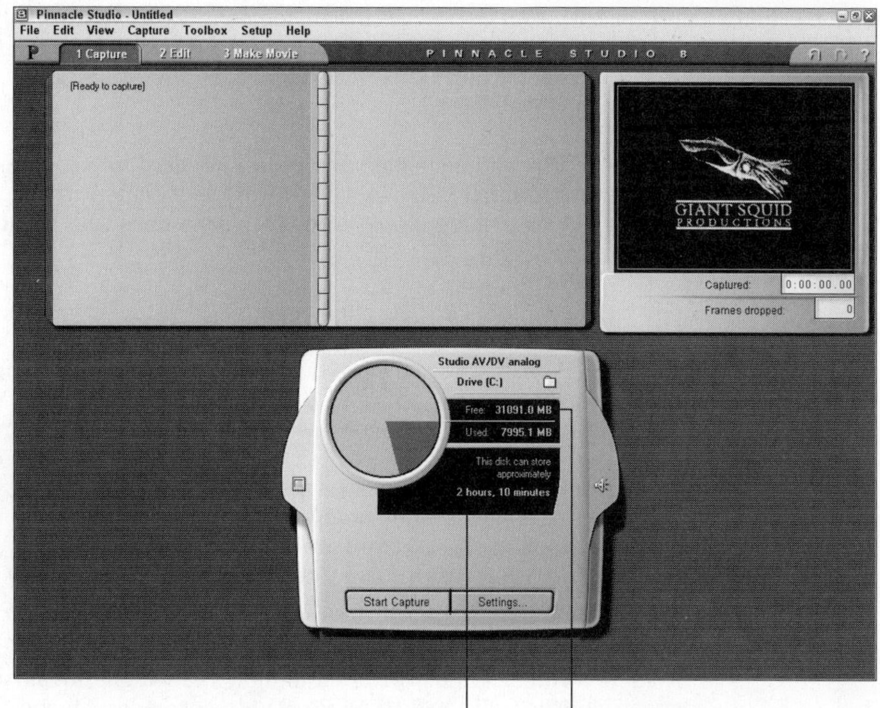

Figure 6-1: Most capture programs provide an estimate of how much video you can capture.

Using a capture card

The best quality in analog video capture is available if you use a special video-capture card. Analog-capture cards are available at many computer and electronics retailers. A capture card actually connects to the motherboard inside your computer — so installation will require some expertise in working with computer hardware. Also, make sure that your computer actually has room to add an expansion card. In Chapter 2, I briefly discuss how to identify empty expansion slots and install a FireWire card in a Windows PC. Installing an analog capture card is very similar.

If you buy a capture card, make sure it can capture analog video. Many FireWire cards are marketed as digital-video capture cards, but if you don't read the packaging carefully, you might be confused about the card's capabilities. If a FireWire card doesn't *specifically* say that it can *also* capture analog video, assume that it can't.

I use a capture card made by Pinnacle Systems (`www.pinnaclesys.com`) called the Pinnacle AVDV Capture Card. This card has FireWire ports for capturing digital video, as well as an external breakout box with analog connectors for capturing analog video. Most analog capture cards have breakout boxes because there usually isn't enough room on the back of a narrow expansion card for all the necessary analog connectors. The AVDV card comes the Pinnacle Studio Deluxe package, which retails for $300. If your PC already has a FireWire port, look into Pinnacle Studio AV ($130), which comes with an analog-capture card.

If you decide to buy and install an analog-capture card in your computer, I strongly recommend that you use the software that came with the card when you're ready to capture. A capture card normally comes with software that is optimized for that card.

After you have a capture card installed in your computer, all you need to do is connect your VCR or analog camcorder to the appropriate ports. Usually those ports are located on an external breakout box (see Figure 2-7 in Chapter 2) because there probably isn't room on the back of the actual card. You will probably have to choose from among several different kinds of connectors:

- **Composite:** Composite connectors — the most common type — are often used to connect video components in a home entertainment system. Composite connectors are also sometimes called *RCA jacks* and use only one connector for the video signal. A composite video connector is usually color-coded yellow. Red and white composite connectors are for audio. Make sure you connect all three.
- **S-Video:** S-video connectors are found on many higher-quality analog camcorders as well as S-VHS VCRs. S-Video provides a higher-quality picture, so use it if it's available as an option. The S-Video connector only carries video, so you'll still need to use the red and white audio connectors for sound.
- **Component:** Component video connectors often look like composite connectors, but the video image is broken up over three separate connectors color-coded red, green, and blue. The red cable is sometimes also labeled R-Y and carries the red portion of the video image, minus brightness information. The green cable — sometimes labeled Y — carries brightness information. Video geeks like to say *luminance* instead of brightness, but it means the same thing. The blue wire — sometimes labeled B-Y — carried the blue portion of the image, minus brightness. Component video provides a higher quality video image, but it's usually only found on the most expensive, professional-grade video capture cards. Like S-Video, component connectors don't carry audio, so you'll still need to hook up the red and white audio cables.

When you're done hooking everything up, you'll probably have quite a rat's nest of cables going everywhere. Your capture software won't be able to control your VCR or camcorder through the analog cables, so you'll have to manually press Play on the device before you can start capturing in the software.

Using a video converter

Video converters are kind of neat because they don't require you to break out the tools and open up your computer case. As their name implies, video converters *convert* analog video to digital before it even gets inside your computer. The converter has connectors for your analog VCR or camcorder, and it connects to your computer via the FireWire port. Converters are available at many electronics retailers for about $200. Common video converters include

- **Canopus ADVC-50:** `www.canopuscorp.com`
- **Data Video DAC-100:** `www.datavideo-tek.com`
- **Dazzle Hollywood DV Bridge:** `www.dazzle.com`

All three of these video converters are compatible with both Macintosh and Windows computers. Video converters pipe in video using a FireWire port — so as far as your computer is concerned, you're capturing digital video. Capturing analog video using a video converter is just like capturing digital video (a process described in Chapter 5). The only difference is that you'll have to manually press Play on your analog camcorder or VCR before you start to capture.

If you have a digital camcorder, you might be able to use it as a video converter for analog video. Digital camcorders have analog connectors so you can hook them up to a regular VCR. Hook up a VCR to the camcorder, set up the camcorder and your computer for digital capture, and push play on the VCR. If the video picture from the VCR appears on the capture-preview screen on your computer, you can use this setup to capture analog video. This should work without having to first record the video onto tape in your digital camcorder, which means that using your camcorder as a converter won't cause increased wear on the camcorder. Alas, some camcorders won't allow analog input while a FireWire cable is plugged in. You'll have to experiment with your own camera to find out if this method will work.

Capturing Video

After you have your capture hardware set up and connected properly, you're ready to start capturing video. If you're using an external video converter connected to your computer's FireWire port, follow the instructions in Chapter 5 for digital capture. The only difference is you'll have to manually press Play on the analog VCR or camcorder before you can start capturing.

Apple iMovie can't capture video from an analog capture card. To capture analog video in iMovie, you must use a video converter connected to your Mac's FireWire port. Capturing video in this manner is just like capturing digital video, as described in Chapter 5.

The exact procedure for capturing analog video varies; it depends on the software you're using. Because I feature Pinnacle Studio throughout this book, I'll show that program here for the sake of consistency. Begin by opening Studio, and then choose View⇨Capture. Studio's capture window appears, as shown in Figure 6-1. The next section shows you how to adjust Studio settings to get ready for analog capture.

Adjusting video-capture settings in Studio

To make sure Studio is ready to capture analog video instead of digital video, click the Settings button at the bottom of the capture window. The Pinnacle Studio Setup Options dialog box appears, as shown in Figure 6-2. Click the Capture Source tab to bring it to the front. On that tab, review the following settings:

- **Video:** Choose your analog capture source in this menu. The choices in the menu will vary depending upon what hardware you have installed.
- **Audio:** This menu should match the Video menu.
- **Use Overlay and Capture Preview:** I recommend that you leave both these options disabled. Leaving these options on may cause dropped frames (that is, some video frames are missed during the capture) on some computers.
- **Scene detection during video capture:** This setting determines when a new scene is created. For most purposes, I recommend choosing "Automatic based on video content" for analog capture.
- **Data rate:** This section tells you how fast your hard disk can read or write. Click the Test Data Rate button to get a current speed estimate. Ideally, both numbers should be higher than 10,000 kilobytes per second for analog capture. If your system can't reach that speed, see Chapter 5 for information on what you can do to speed up your hard disk.

If you have trouble with dropped frames when you capture analog video, try disabling automatic scene detection by choosing the No Automatic Scene Detection option. Scene detection uses some computer resources, and it can cause some dropped frames if your computer isn't quite fast enough.

When you're done reviewing settings on the Capture Source tab, click OK to close the Setup Options dialog box. You are almost ready to begin capturing analog video. Like many analog video capture programs, Studio lets you fine-tune the audio and video that will be captured. Click the buttons on either side of the capture controller to open the Video Input and Audio Capture control panels as shown in Figure 6-3. These control panels allow you to adjust color, brightness, and audio levels of the incoming video.

On the Video Input control panel, first choose whether you're going to capture video from the Composite or S-Video connectors using the radio buttons under Video Input. The Audio Capture control panel allows you to turn audio capture on or off.

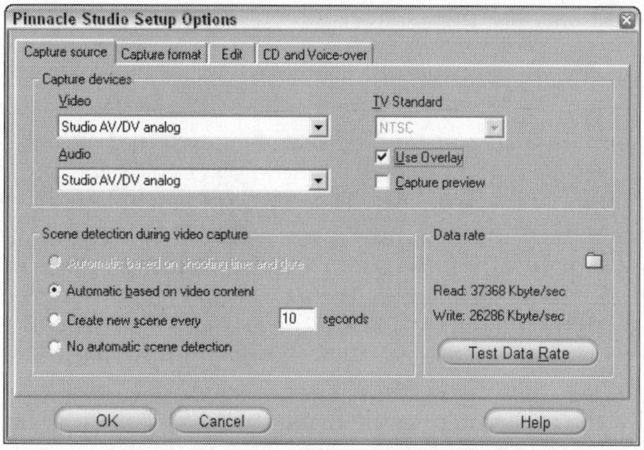

Pinnacle Studio Setup Options

Capture source | Capture format | Edit | CD and Voice-over

Capture devices

Video: Studio AV/DV analog

Audio: Studio AV/DV analog

TV Standard: NTSC

Use Overlay

Capture preview

Scene detection during video capture

Automatic based on shooting time and date

Automatic based on video content

Create new scene every 10 seconds

No automatic scene detection

Data rate

Read: 37368 Kbyte/sec

Write: 26286 Kbyte/sec

Test Data Rate

OK | Cancel | Help

Figure 6-2: Choose your analog capture source here.

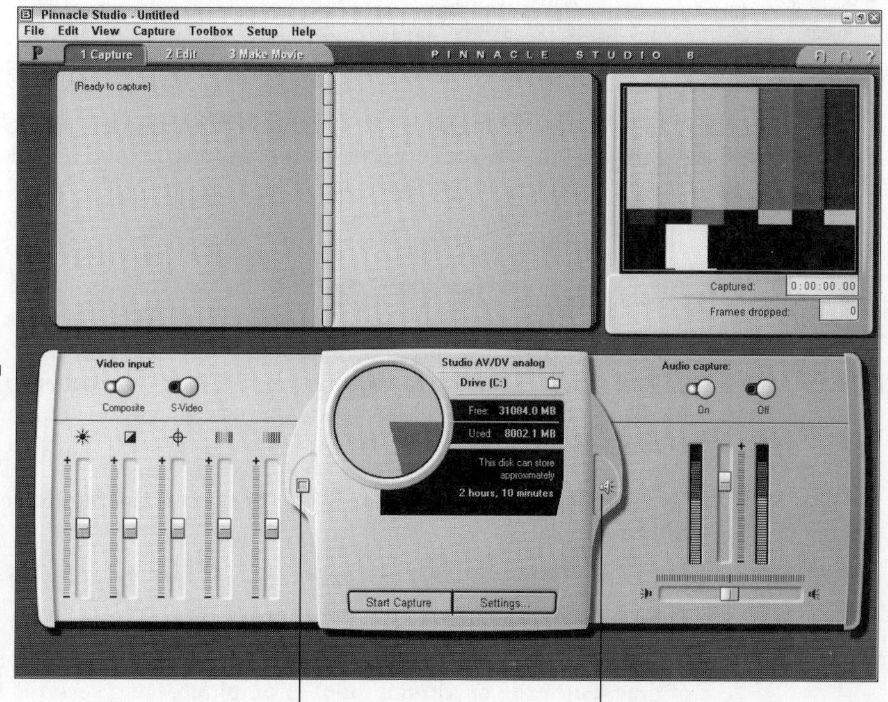

Figure 6-3: Use these control panels to fine-tune your incoming video and audio.

Now press Play on your VCR or camcorder to begin playing the video you plan to capture, but don't start capturing it yet. As you play the analog video, watch the picture in the preview screen in the upper right corner of the Studio program window. If you don't like the picture quality, you can use the brightness, contrast, sharpness, hue, and color saturation sliders on the Video Input control panel to adjust the picture. Experiment a bit with the settings to achieve the best result.

As you play your analog tape, you'll probably notice that although you can see the video picture in the preview window, you can't hear the sound through your computer's speakers. That's okay. Keep an eye on the audio meters on the Audio Capture control panel. They should move up and down as the movie plays. Ideally, most audio will be in the high green or low yellow portion of the audio meters. Adjust the audio-level slider between the meters if the levels seem too high or too low. If the audio seems biased too much to the left or right, adjust the balance slider at the bottom of the control panel.

If you're lucky, the tape you're going to capture from has color bars and a tone at the beginning or end (as shown in Figure 6-3). In that case, use the bars and tone to calibrate your video picture and sound. The tone is really handy because it's a standard 1KHz (kilohertz) tone designed specifically for calibrating audio levels. Adjust the audio-level slider so both meters read just at the bottom of the yellow, as shown in Figure 6-3.

When you're done playing with the settings in the Video and Audio control panels, stop the tape in your VCR or camcorder and rewind it back to where you want to begin capturing.

Capturing your video

Once all your settings are, uh, *set*, you're ready to start capturing. (Finally!) I recommend that you rewind the tape in your VCR or camcorder to at least 15 seconds *before* the point at which you want to begin capturing. Then follow these steps:

1. **Click the Start Capture button at the bottom of the Studio capture window.**

   The Capture Video dialog box appears, as shown in Figure 6-4.

2. **Enter a descriptive name for the capture.**

3. **If you want to automatically stop capturing after a certain period of time, enter the maximum number of minutes and seconds for the capture.**

   In Figure 6-4, you can see that I want to capture only six minutes of video.

4. **Press Play on the VCR or camcorder.**

5. **Click Start Capture within the Capture Video dialog box.**

The capture process begins, and you'll notice that the green Start Capture button changes to the red Stop Capture button. As Studio captures your video, keep an eye on the Frames Dropped field under the preview window. If any frames are dropped, try to determine the cause and then recapture the video. (Chapter 5 suggests some things you can do to prevent dropped frames.) Common causes of dropped frames include programs running in the background, power saver modes, or a hard disk that hasn't been defragmented recently.

6. **Click Stop Capture when you're done capturing.**

   Studio reviews the video that has been captured and improves scene detection if possible. When the process is done, the captured clips appear in Studio's clip album.

Figure 6-4: Choose your analog-capture source here.

Chapter 7

Importing Audio

In This Chapter

- Audio fundamentals
- Recording audio
- Working with CD audio
- Using MP3 files

We often think of movies as a purely visual medium. As a result, overlooking audio is easy. But most video and film professionals will tell you that audio is almost as — if not more — important than the visual picture. Those same pros will probably tell you that audiences can forgive or ignore a few visual flaws, but poor-quality audio immediately enhances the cheese factor of a movie.

Movie sound is a pretty big subject, which is why I've devoted not one but *two* chapters in this book to audio. This chapter helps you understand the fundamentals of audio, and it helps you obtain and record better quality audio. After you have some audio source material to work with, check out Chapter 10 for more on working with audio in the movie editing software on your computer.

Understanding Audio

Consider how audio affects the feel of a video program. Honking car horns on a busy street; crashing surf and calling seagulls at a beach; a howling wolf on the moors — these sounds help us identify a place as quickly as our eyes can, if not quicker. If a picture is worth a thousand words, sometimes a sound in your movie is worth a thousand pictures.

What is audio? Well, if I check my notes from high-school science class, I get the impression that audio is produced by sound waves moving through the air. Human beings hear those sound waves when they make our eardrums vibrate. The speed at which a sound makes the eardrum vibrate is the *frequency*. Frequency is measured in kilohertz (kHz), and one kHz equals one thousand vibrations per second. (You could say it really *hertz* when your eardrums vibrate . . . get it?) A lower-frequency sound is perceived as a lower pitch or tone, and a higher-frequency sound is perceived as a high pitch or tone. The volume or intensity of audio is measured in *decibels* (dB).

Understanding sampling rates

For over a century, humans have been using analog devices (ranging from wax cylinders to magnetic tapes) to record sound waves. As with video, digital audio recordings are all the rage today. Because a digital recording can only contain specific values, it can only approximate a continuous wave of sound; a digital recording device must "sample" a sound many times per second; the more samples per second, the more closely the recording can approximate the live sound (although a digital approximation of a "wave" actually looks more like the stairs on an Aztec pyramid). The number of samples per second is called the *sampling rate*. As you might expect, a higher sampling rate provides better recording quality. CD audio typically has a sampling rate of 44.1 kHz — that's 44,100 samples per second — and most digital camcorders can record at a sampling rate of 48kHz. You will work with sampling rate when you adjust the settings on your camcorder, import audio into your computer, and export movie projects when they're done.

Delving into bit depth

Another term you'll hear bandied about in audio editing is *bit depth*. The quality of a digital audio recording is affected by the number of samples per second, as well as by how much information each sample contains. The amount of information that can be recorded per sample is the bit depth. More bits per sample mean more information — and generally richer sound.

Many digital recorders and camcorders offer a choice between 12-bit and 16-bit audio; choose the 16-bit setting whenever possible. For some reason, many digital camcorders come from the factory set to record 12-bit audio. There is no advantage to using the lower setting, so always check your camcorder's documentation and adjust the audio-recording bit depth up to 16-bit if it isn't there already.

Recording Audio

At some point, you'll probably want to record some narration or other sound to go along with your movie project. Recording great-quality audio is no simple matter. Professional recording studios spend thousands or even millions of dollars to set up acoustically superior sound rooms. I'm guessing you don't have that kind of budgetary firepower handy, but if you're recording your own sound, you can get nearly pro-sounding results if you follow these basic tips:

- **Use an external microphone whenever possible.** The built-in microphones in modern camcorders have improved greatly in recent years, but they still present problems. They often record undesired ambient sound near the camcorder (such as audience members) or even mechanical sound from the camcorder's tape drive. If possible, connect an external microphone to the camcorder's mic input.
- **Eliminate unwanted noise sources.** If you *must* use the camcorder's built-in mic, be aware of your movements and other things that can cause loud, distracting noises on tape. Problem items can include a loose lens cap banging around, your finger rubbing against the mic, wind blowing across the mic, and the *swish-swish* of those nylon workout pants you wore this morning.
- **Control ambient noise.** True silence is a very rare thing in modern life. Before you start recording audio, carefully observe various sources of noise. These could include your neighbor's lawn mower, someone watching TV in another room, extra computers, and even the heating duct from your furnace or air conditioner. Noise from any (or all) of these things can reduce the quality of your recording.
- **Try to minimize sound reflection.** Audio waves reflect off any hard surface, which can cause echoing in a recording. Cover the walls, floor, and other hard surfaces with blankets to reduce sound reflection.
- **Obtain and use a high-quality microphone.** A good mic isn't cheap, but it can make a huge difference in recording quality.
- **Watch for trip hazards!** In your haste to record great sound, don't forget that your microphone cables can become a hazard on-scene. Not only is this a safety hazard to anyone walking by, but if someone snags a cable, your equipment could be damaged as well. If necessary, bring along some duct tape to temporarily cover cables that run across the floor.

The easiest way to record audio is with a microphone connected to your computer, although some computers can make a lot of noise with their whirring hard disks, spinning fans, and buzzing monitors. The following sections show you how to record audio using a microphone connected to the microphone jack on your computer.

Recording audio with your Macintosh

You can record audio directly in iMovie. In fact, if you have an iMac, PowerBook, or newer iBook, your computer already has a built-in microphone. It will work for very basic narration, but keep in mind that the built-in mic does not record studio-quality sound. If you have a better mic, connect it to the computer's external microphone jack. The following sections show you how to set up your microphone and record audio.

Connecting a tape player to your computer

If you recorded audio onto an audio tape, you can connect the tape player directly to your computer and record from it just as if you were recording from a microphone. Buy a patch cable (available at almost any electronics store) with two male mini-jacks (standard small audio connectors used by most current headphones, microphones, and computer speakers). Connect one end of the patch cable to the headphone jack on the tape player and connect the other end to your computer's microphone jack. The key, of course, is to have the right kind of cable. You can even buy cables that allow you to connect an old record player to your computer and record audio from your old LPs.

Once connected, you can record audio from the tape using the same steps described in the sections in this chapter on microphone recording. You'll have to coordinate your fingers so that you press Play on the tape player and click Record in the recording software at about the same time. If the recording levels are too high or too low, adjust the volume control on the tape player.

Setting up an external microphone

Some microphones can connect to the USB port on your Mac. A USB microphone will be easier to use because your Mac will automatically recognize it and select the USB mic as your primary recording source. If your external microphone connects to the regular analog microphone jack — and your Mac already has a built-in mic — you may find that iMovie doesn't recognize your external microphone. To correct this problem, you must adjust your system's Sound settings:

1. **Open the System Preferences window by choosing Apple⇨System Preferences.**
2. **Double-click the Sound icon to open the Sound preferences dialog box.**
3. **Click the Input tab to bring it to the front.**
4. **Open the Microphone pull-down menu and choose External Microphone/Line In.**
5. **Press ⌘+Q to close the Sound dialog box and System Preferences.**

Your external microphone should now be configured for use in iMovie.

Recording in iMovie

Once you've decided which microphone to use and you've configured it as described in the previous section, you're ready to record audio using iMovie. As with most tasks in iMovie, recording is pretty easy:

1. **Open the project for which you want to record narration or other sounds, and switch to the timeline view if you're not there already.**

For more on editing in the timeline, see Chapter 8. If you don't have a project yet and just want to record some audio, that's okay too.

2. **If you're working with a current movie project, move the play head to the spot where you want to begin recording.**
3. **Click the Audio button above the timeline to open the audio pane (as shown in Figure 7-1).**
4. **Say a few words to test the audio levels.**

   As your microphone picks up sound, the audio meter in iMovie should indicate the recording level. If the meter doesn't move at all, your microphone probably isn't working. You'll notice that as sound levels rise, the meter changes from green to yellow and finally red. For best results, try to keep the sound levels close to the yellow part of the meter.

   If the audio levels are too low, the recording may have a lot of unwanted noise relative to the recorded voice or sound. If levels are too high, the audio recording could pop and sound distorted. Unfortunately, iMovie doesn't offer an audio level adjustment for audio you record, so you'll have to fine-tune levels the old fashioned way, by changing the distance between the microphone and your subject.
5. **Click Record and begin your narration.**

   The movie project plays as you recite your narration.
6. **When you're done, click Stop.**

   An audio clip of your narration appears in the timeline, as shown in Figure 7-1.

For more on working with audio clips in your movie projects, see Chapter 10.

Recording voice-over tracks in Pinnacle Studio

Recording audio in Windows is pretty easy. Most video-editing programs — including Pinnacle Studio and Windows Movie Maker — give you the capability to record audio directly in the software. Before you can record audio, however, your computer must have a sound card and a microphone. (If your computer has speakers, it has a sound card.) The sound card should have a connector for a microphone as well. Check the documentation for your computer if you can't find the microphone connector.

After your hardware is set up correctly, you're ready to record audio in Pinnacle Studio or most any other movie making program. To record audio in Studio, follow these steps:

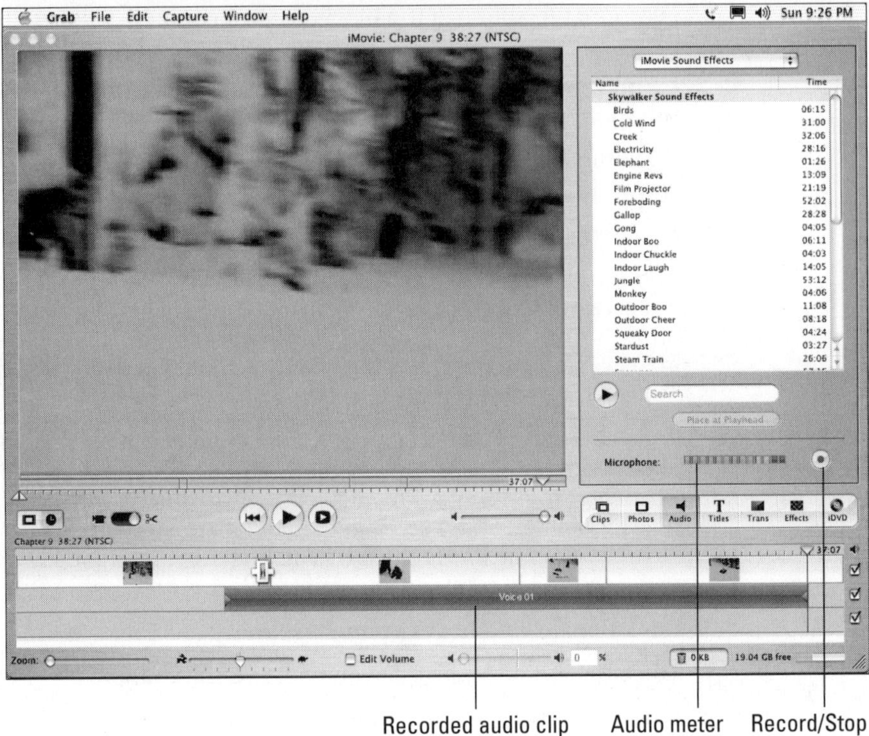

Figure 7-1: You can record audio directly in iMovie.

1. **Open the movie project for which you want to record audio, and switch to the timeline view if you aren't there already.**

   For more on editing in the timeline, see Chapter 8. (If you don't have a project yet and just want to record some audio, that's okay too.)

2. **If you're working with a current movie project, move the play head to the spot where you want to begin recording.**

3. **Choose Toolbox⇨Record Voice-over.**

   The voice-over recording studio appears, as shown in Figure 7-2.

4. **Say a few words to test the audio levels.**

   As your microphone picks up sound, the audio meter in Studio should indicate the recording level. If the meter doesn't move at all, your microphone probably isn't working. You'll notice that as sound levels rise, the meter changes from green to yellow and finally red. For best results, try to keep the sound levels in or near the yellow part of the meter. You can fine-tune the levels by adjusting the Recording Volume slider.

5. **Click Record.**

   A visible three-second countdown appears in the recording-studio window, giving you a couple of seconds to get ready.

6. **When recording begins and your movie project starts to play, recite your narration.**
7. **When you're done, click Stop.**

   An audio clip of your narration appears in the timeline, as shown in Figure 7-2.

For more on working with audio clips in your movie projects, see Chapter 10.

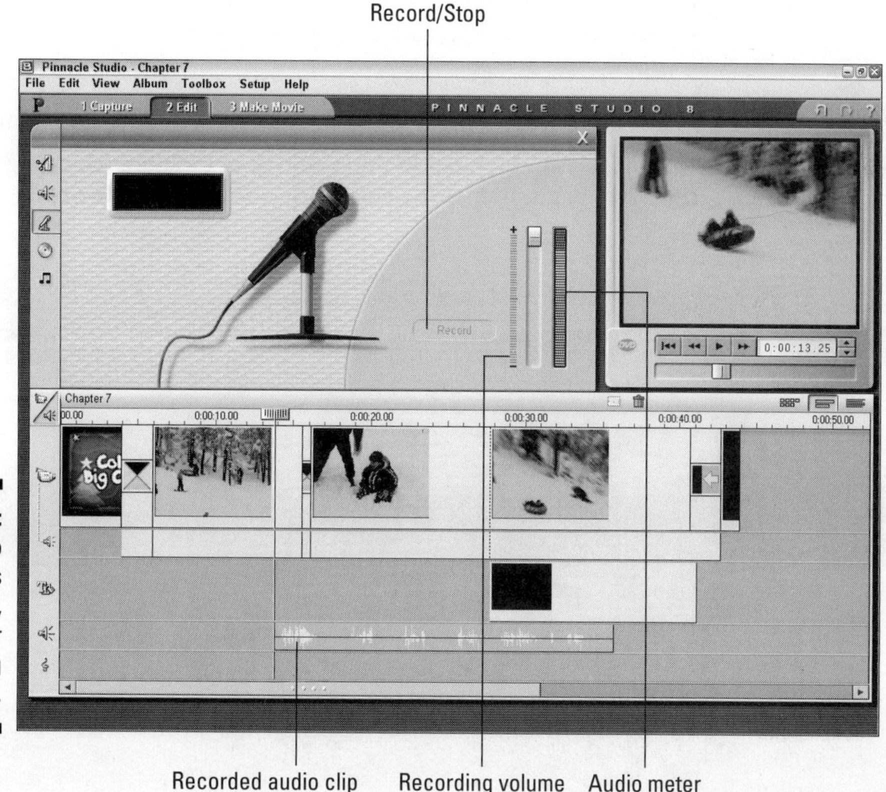

Figure 7-2: Studio includes its own, er, *studio* for recording audio.

Working with CD Audio

If you're even remotely interested in technology — and because you're reading this book, I'll go out on a limb and guess that you are — you probably started upgrading your music collection to compact discs (CDs) many years ago. At some point, you'll probably want to use some of the music on those CDs for your movie soundtracks. Most editing programs make it pretty easy to use CD audio in your movie projects. The next couple of sections show you how to use CD audio in Pinnacle Studio and iMovie.

Importing CD audio in iMovie

If you're using Apple iMovie, you can take audio directly from your iTunes library or import audio from an audio CD. Here's the basic drill:

1. **Put an audio CD in your CD-ROM drive, and then click the Audio button above the timeline on the right side of the iMovie screen.**

   The audio pane appears, as shown in Figure 7-3.

2. **If iMovie Sound Effects are currently shown, open the pull-down menu at the top of the audio pane and choose Audio CD.**

   A list of audio tracks appears, as shown in Figure 7-3.

3. **Select a track and click Play in the audio pane to preview the song.**
4. **Click Place at Playhead in the Audio panel to place the song in the timeline, beginning at the current position of the play head.**

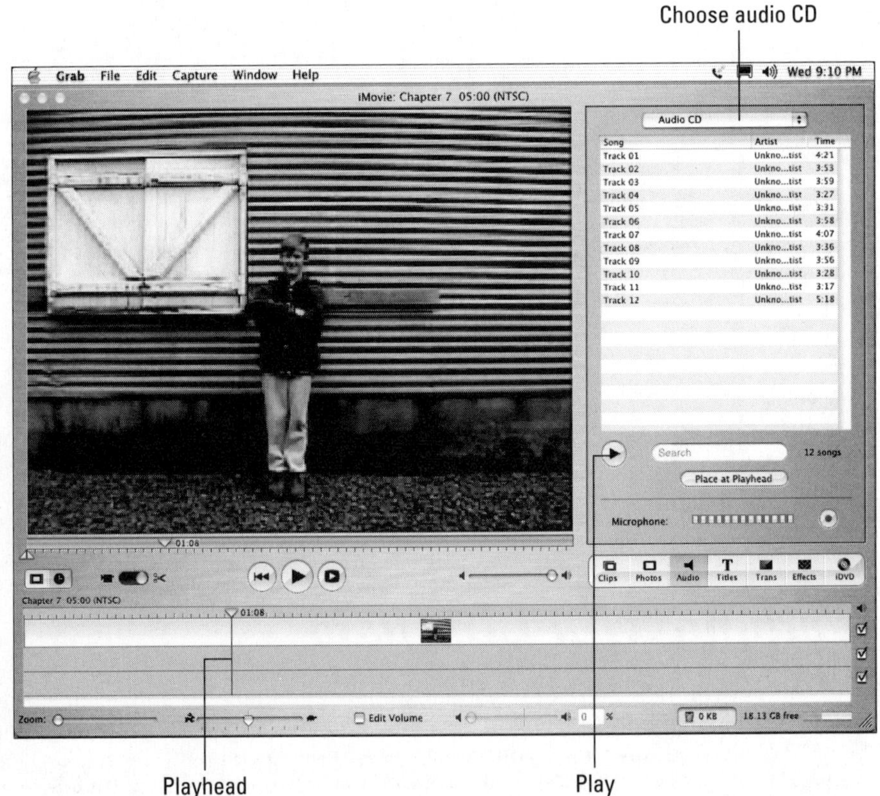

Figure 7-3: Songs from audio CDs can be imported directly into your movie projects using iMovie.

Importing CD audio in Studio

Pinnacle Studio provides an audio toolbox to help you work with CD audio and other formats. To open the audio toolbox in Studio, place an audio CD in your CD-ROM drive and choose Toolbox⇨Add CD Music. You may be asked to enter a title for the CD. The exact title doesn't matter, so long as it's something you will be able to identify and remember later. The audio toolbox will appear as shown in Figure 7-4. The audio toolbox allows you to perform a variety of actions:

- Choose audio tracks from the CD using the Track menu.
- Use playback controls in the audio toolbox to preview audio tracks.
- Add only a portion of the audio clip to your movie by adjusting the in point and out point markers.
- Add a track to the movie as a clip in the background music track at the bottom of the timeline (as in Figure 7-4) by clicking the Add to Movie. The clip is at the current position of the play head. (See Chapter 10 for more on editing audio in your movie projects.)

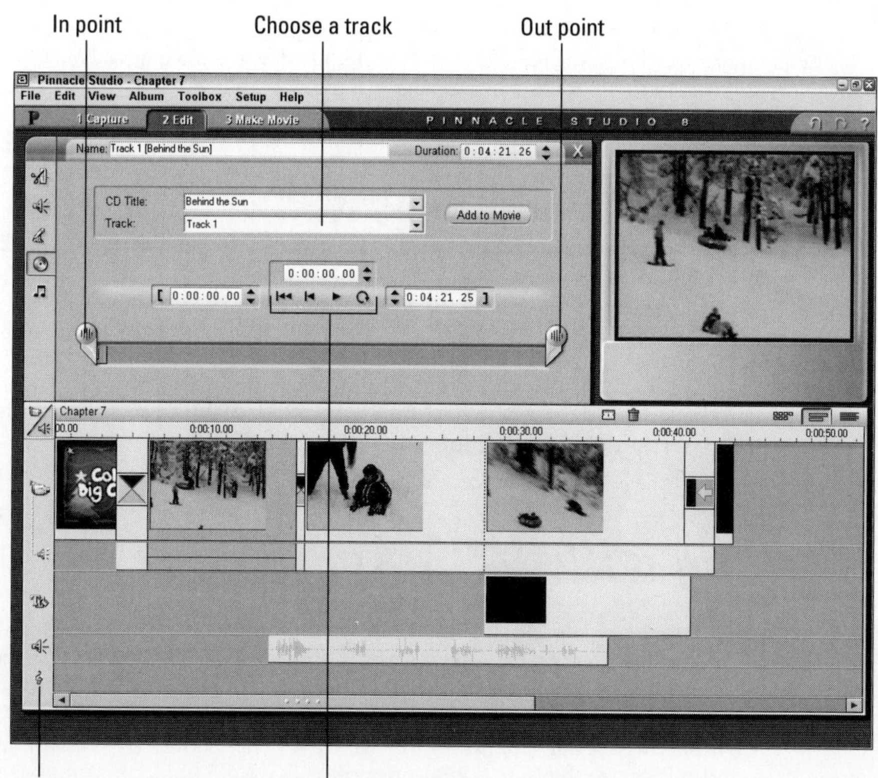

Figure 7-4: Use Studio's audio toolbox to add CD audio to your movie.

Comprehending copyrights

As I show throughout this chapter, music can easily be imported into your computer for use in your movies. The tricky part is obtaining the rights to use that music legally. Realistically, if you're making a video of your daughter's birthday party, and you only plan to share that video with a grandparent or two, Kool and the Gang probably don't care if you use the song "Celebration" as a musical soundtrack. But if you start distributing that movie all over the Internet — or (even worse) start selling it — you could have a problem that involves the band, the record company, and lots and lots of lawyers.

The key to using music legally is licensing. You can license the right to use just about any music you want, but it can get expensive. Licensing "Celebration" for your movie project (for example) could cost hundreds of dollars or more. Fortunately, more affordable alternatives exist. Numerous companies offer CDs and online libraries of stock audio and music, which you can license for as little as $30 or $40. You can find these resources by searching the Web for "royalty free music," or visit a site such as `www.royaltyfree.com` or `www.royaltyfreemusic.com`. You usually must pay a fee to download or purchase the music initially, but after you have purchased it, you can use the music however you'd like. If you use audio from such a resource, make sure you read the licensing agreement carefully. Even though the music is called "royalty free," you still may be restricted on how many copies you may distribute, how much money you can charge, or what formats you may offer.

Another — more affordable — alternative may be to use the stock audio that comes with some moviemaking software. Pinnacle Studio, for example, comes with a tool called SmartSound which automatically generates music in a variety of styles and moods. (See Chapter 10 for more on working with SmartSound audio.) Although you should always carefully review the software license agreements to be sure, normally the audio that comes with moviemaking software can be used in your movie projects free of royalties.

After you add a CD audio track to your movie project, play the project to preview your addition. The first time you preview the movie with the CD audio track added, Studio captures the required audio from the CD. If the disc isn't in the CD-ROM drive the first time you try to play the project, Studio asks you to insert the disc before the process can continue.

Working with MP3 Audio

MP3 is one of the most common formats for sharing audio recordings today. MP3 is short for *MPEG Layer-3*, and MPEG is short for *Motion Picture Experts Group*, so really you can think of MP3 as an abbreviation of an abbreviation. I'm sure that in a few years an MP3 file will simply be called an "M" or "P" (or maybe even a "3") file — but whatever their collective nickname, MP3 audio files are likely to remain popular. The MP3 file format makes for very small files — you can easily store a lot of music on a hard drive or CD — and those files are easy to transfer over the Internet.

Who am I kidding? You probably already know all about MP3 files. You might even have some MP3 files already stored on your computer. If so, using those MP3 files for background music in your movie projects is really easy:

- **In iMovie:** Pull MP3 files directly from your iTunes library into iMovie, using the procedure described earlier in this chapter for importing CD Audio. Simply choose iTunes from the pull-down menu at the top of the audio pane.
- **In Studio:** Choose Album⇨Sound Effects to show the sound-effects album. Click the folder icon and browse to the folder on your hard drive that contains the MP3 files you want to use (as shown in Figure 7-5). When a list of MP3 files appears in the album, simply drag-and-drop them on the background music track of your timeline.

Storing audio on your hard disk is handy because the audio will be easier to plop into your movie projects. MP3 is a great format to use because the audio sounds about as good as CD audio, but it takes up a lot less storage space. If you're not sure how to copy music from audio CDs onto your hard disk in MP3 format — the process of converting audio to the MP3 format is often called *encoding* or *ripping* — check out the next two sections.

Ripping MP3 files on a Mac

The process of turning an audio file into an MP3 file is sometimes called *ripping* or *encoding*. Apple has thoughtfully provided the capability to create MP3 files with its free audio-library-and-player program, iTunes. To download the latest version of iTunes, visit `www.apple.com/itunes/` and follow the instructions there. After iTunes is installed on your computer, copying audio onto your hard drive in MP3 format is quite simple:

1. **Insert an audio CD into your CD-ROM drive.**
2. **If iTunes doesn't launch automatically, open the program using the Dock or your Applications folder.**
3. **With the iTunes program window active as shown in Figure 7-6, choose iTunes⇨Preferences.**

   The iTunes Preferences dialog box opens.
4. **Click the Importing button at the top of the Preferences dialog box.**
5. **Make sure that MP3 Encoder is selected in the Import Using menu, and then click OK.**

   The iTunes Preferences dialog box closes and you are returned to the main iTunes window.

Figure 7-5: Access MP3 audio files through Studio's sound-effects album.

6. **Place check marks next to the songs you want to import.**

   You can use the playback controls in the upper-left corner of the iTunes window to preview tracks. In Figure 7-6, for example, I have chosen three tracks to import.

7. **Click Import in the upper right corner of the iTunes screen.**

   The songs are imported; the process may take several minutes. When it's done, the imported songs are available through your iTunes library for use in iMovie projects.

Ripping MP3 files in Windows

As I mention earlier, the process of turning audio files into MP3 files is sometimes called *encoding* or *ripping*. Microsoft provides a free audio-player program called Windows Media Player — WMP for short. It comes with Windows, and you can download the latest version from www.windowsmedia.com. Like Apple's iTunes for the Macintosh, WMP allows you to copy music from audio CDs to your hard drive in a high-quality (yet compact) format. Unfortunately, as delivered, WMP does not rip files in MP3 format. Instead, it uses the Windows Media Audio (WMA) format.

Figure 7-6: iTunes can rip CD audio onto your hard disk in MP3 format.

Windows Media files are about as small as MP3 files, but it's a proprietary format: Most video-editing programs (including Pinnacle Studio) cannot import WMA files directly. If you want to import music from CDs into a Studio movie project, it's better to copy the music directly from within Studio, as described earlier in this chapter.

If you really want to be able to copy music onto your hard drive in MP3 format, you'll have to obtain commercially available MP3 encoding software. Such programs are available at most electronics stores, and you can also download software from Web sites such as www.tucows.com.

I use a tool called CinePlayer from Sonic Solutions (www.cineplayer.com). This $20 tool works as a plug-in for Windows Media Player in Windows XP, and allows WMP to both encode MP3 files and play DVD movies. After it's installed, I simply open Windows Media Player and choose Tools➪Options. Then, on the Copy Music tab of the Options dialog box (shown in Figure 7-7), I can choose MPEG Layer-3 Audio in the Format drop-down box, and adjust quality settings as I see fit. The MPEG Layer-3 option is available here only because I have the CinePlayer plug-in installed. With these settings, WMP uses the MP3 format instead of WMA when I copy music to my hard disk.

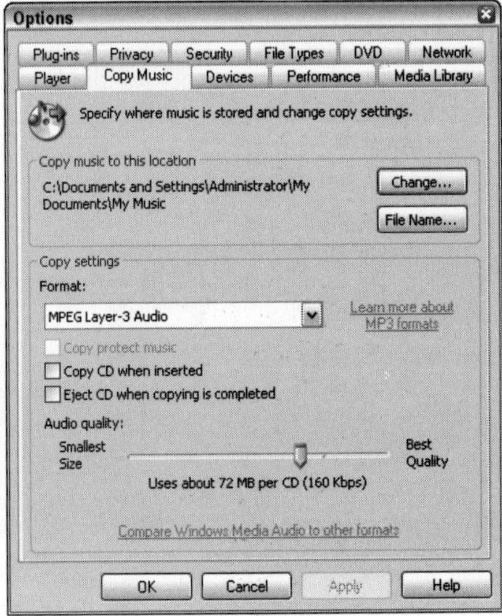

Figure 7-7: The CinePlayer plug-in allows me to copy music in MP3 format using Windows Media Player.

Part III
Editing Your Movie

In this part...

Perhaps the best thing about digital video is the ease with which you can edit it. Just connect your digital camcorder to your computer's FireWire port, import the video (as described in Part II), and you're ready to edit.

When you edit video, you put scenes in the order you want them, remove unwanted material, add titles (words on the screen titles), and create special, visually appealing transitions between video clips. You can also add sound effects, a music soundtrack, and some still pictures or graphics if you like. And of course, special effects can add a lot of excitement to your movie as well. This part shows you how to do these things using software that is available to you right now.

Chapter 8

Hollywood Comes Home: Basic Editing

In This Chapter

- Starting an editing project
- Working with video clips
- Assembling a movie project
- Fine-tuning the project

By themselves, digital video cameras aren't *that* big a deal. Sure, digital camcorders offer higher quality, but the video quality of analog Hi8 camcorders really wasn't too bad. Digital video doesn't suffer from generational loss (where some video quality is lost each time the video is copied or even played) like analog video does, but again, this isn't something that should motivate millions of people to instantly trash their old analog camcorders.

But video editing . . . now, *that's* cool. Until recently, high quality video productions with special effects, on-screen titles, and fancy transitions between scenes were magical productions made by pros using equipment that cost hundreds of thousands (if not millions) of dollars. But thanks to the dual revolutions of digital camcorders and powerful personal computers, all you need for pro-quality video production is a computer, a digital camcorder, a little cable to connect them both, and some software. Within just a few mouse clicks, you'll be making movie magic!

Starting a New Project

Your first step in working on any movie project is to actually create the new project. This step is pretty easy. In fact, when you launch your video-editing program, it will usually start with a new, empty project. But if you need to create a new project for some reason or just want to be sure that you're starting from a clean slate, choose File⇨New Project. No matter if you're using Apple iMovie, Pinnacle Studio, or almost any other video-editing program — a new, empty project window should appear.

After you have a new project started, your first step is usually to capture or import some video so you have something to edit. If you have some video to capture, see Chapter 5 (for digital video) or Chapter 6 (for analog video) for information on how to do that. Captured video will appear as clips in the clips pane (in iMovie) or the Album (in Studio), where it will be ready to use in your movies.

If you don't have any video of your own to import or capture right now, you can still practice editing using sample videos from the CD-ROM that accompanies this book. If you plan to use those sample clips, insert the disc into your CD-ROM drive. To import the sample clips using Pinnacle Studio on a PC, follow these steps:

1. **Click the Edit view mode tab or choose View⇨Edit to ensure that you're in Edit mode.**

   If this is the first time you've used Studio, you'll probably see the sample clips from Pinnacle's Photoshoot sample movie.

2. **Click the Select Video Files button in the clip browser as shown in Figure 8-1.**

   The generic Open File window common to virtually all Windows programs appears.

3. **Browse to the folder** `Samples\Chapter8`**.**

4. **Choose the** `newport` **file and click Open.**

   The file is imported; Studio should automatically detect five scenes in the movie.

If you're using a Mac, import the clips into iMovie by following these steps:

1. **Open iMovie and choose File⇨Import.**

   The generic Open File dialog box common to virtually all OS X programs appears.

2. **Browse to the folder** `Samples\Chapter8` **on the CD-ROM.**

3. **Click-and-drag over all five scene clips to select them and then click Open.**

   The five clips will be imported and will appear in the Clip browser.

The Macintosh-compatible sample files on the CD are in Apple QuickTime format. If you're using Apple iMovie, you need version 3 or higher of iMovie to import QuickTime-format video files. See Appendix C for more on obtaining and installing iMovie 3.

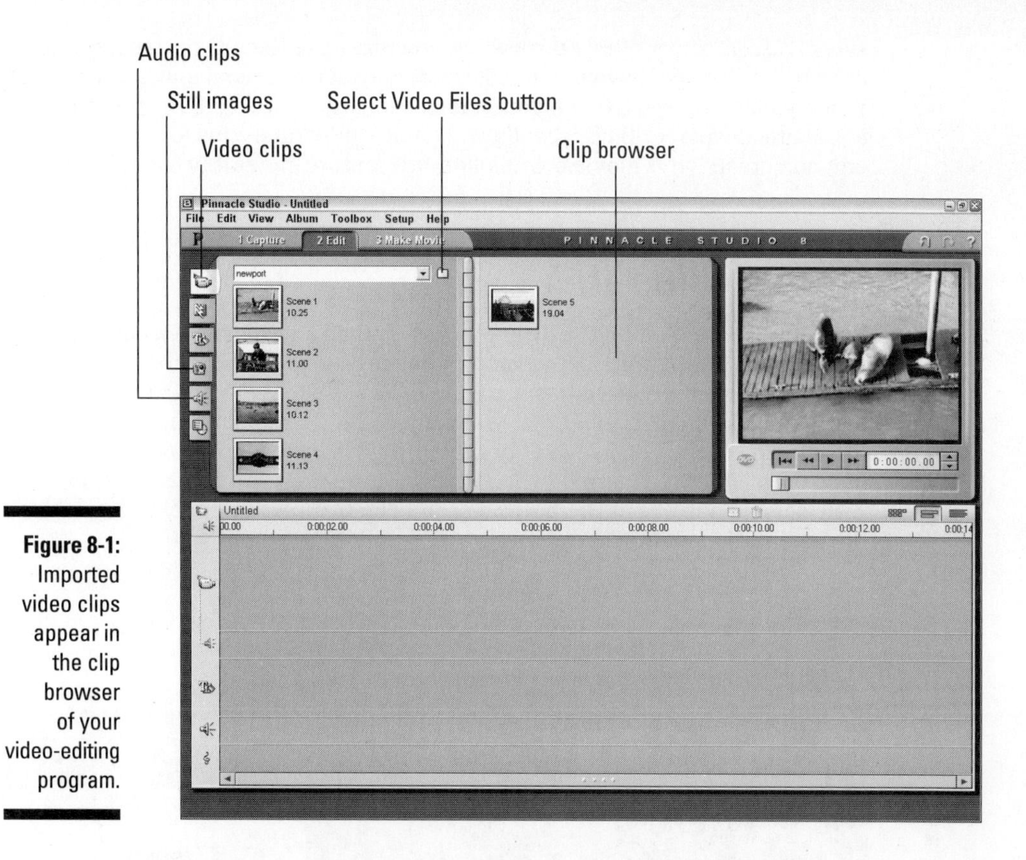

Figure 8-1: Imported video clips appear in the clip browser of your video-editing program.

After you have imported the clips, I recommend that you save and name the project by clicking File➪Save Project. Saving early is important because it not only preserves your work, but it is also required before you can perform certain editing tasks later on.

Working with Clips

As you make movies, you'll quickly find that "clip" is the basic denomination of the media that you work with. You'll spend a lot of time with video clips, audio clips, and even *still* clips (you know, those things that used to be called "pictures" or "photos").

A still clip usually consists of a single picture; an audio clip usually consists of a single song or sound effect; and a video clip usually consists of a single scene. A *scene* most often starts when you press the Record button on your camcorder, and ends when you stop recording again, even if only for a second. When you import video from your camcorder, most video-editing programs automatically detect these scenes and create individual clips for you. As you edit and create your movies, you'll find this feature incredibly useful.

Organizing clips

Virtually all video-editing software stores clips in a grid-like area called a *clip browser* or *album*. Apple iMovie and Pinnacle Studio further subdivide clips by content. Each program has separate browser panels for video clips, audio clips, and still images. In Studio, you can access these panels using tabs along the left side of the clip browser or *album,* as it is called in Pinnacle's documentation (refer to Figure 8-1). In iMovie (Figure 8-2), you can access the panels using buttons at the bottom of the clip browser or *pane,* as Apple's documentation calls it.

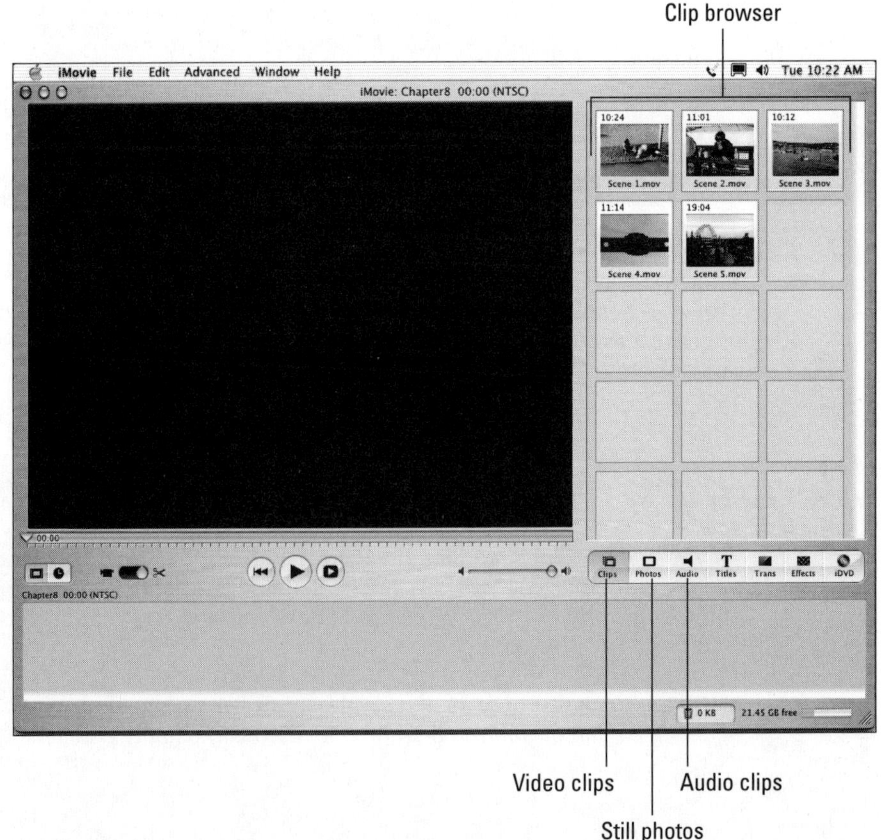

Figure 8-2: The browser helps keep your various clips organized.

The clip browser doesn't just store your clips, it also tells you some important information about them. One of the most important bits of info is the length of the clip. If you're using the sample clips from the CD-ROM, you'll notice the numerals `19.04` or `19:04` next to Scene 5. This tells you that the clip is 19 seconds and four frames long. If you don't see names or lengths listed next to clips in the Studio clip browser, choose Album⇨Details View. This should change the browser view so that it looks more like Figure 8-3.

A video image is actually made up of a series of still images that flash by so quickly that they create the illusion of motion. These still images are called *frames*. Video usually has about 30 frames per second.

The clip also has a name, of course, and you can change the name if you wish. If, for example, you think that "Bridge" would be a more descriptive name than "Scene 5," click the clip once, wait a second, and then click the clip's name once again. You can then type a new name if you want. When you're done typing a new name, just press Enter or click in an empty part of the screen.

If you have a *lot* of clips in the browser, they might not all fit on one screen. In iMovie, simply scroll down (using the scroll bar on the right) to see more clips. In Pinnacle Studio, you can select different groups of clips from the menu at the top of the browser. If there are too many clips in a group to fit on a single page all at once, you can click the arrows (as shown in Figure 8-3) to view additional pages.

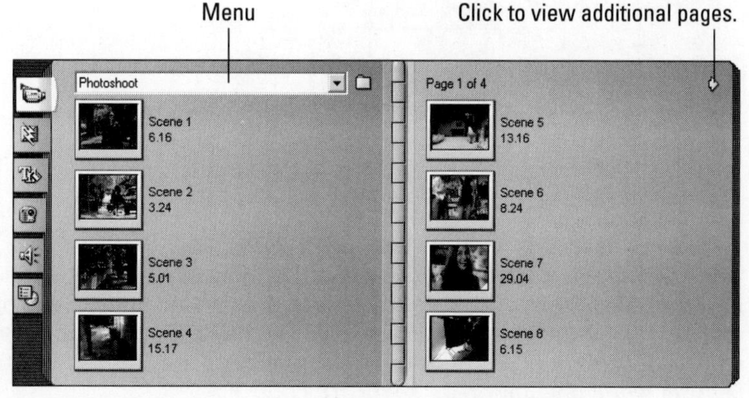

Figure 8-3: In some cases, not all your clips fit on just one page.

In iMovie's clip browser, you can manually rearrange clips by dragging them to new empty blocks in the grid. You can move the clips wherever you want. This handy feature allows you to sort clips on your own terms, rather than just have them listed alphabetically or in some other arbitrary order.

Previewing clips

As you gaze at the clips in your clip browser, you'll notice that a thumbnail image is shown for each clip. This thumbnail usually shows the first frame of the clip; although it may suggest the clip's basic content, you won't really know exactly what the clip contains until you actually preview the whole thing. Previewing a clip is easy: Just click your chosen clip in the browser, and then click the Play button in the playback controls of your software's preview window.

As the clip plays, notice that the play head under the preview window moves. You can move to any portion of a clip by clicking-and-dragging the play head with the mouse. Figure 8-4 shows Studio's preview window and playback controls.

Figure 8-4: Use playback controls and the play head to preview clips.

As you preview clips and identify portions you want to use in your movies, you'll find precise, frame-by-frame control of the play head is crucial. The best way to get that precision is to use keyboard buttons instead of the mouse. Table 8-1 lists important keyboard controls for three popular video-editing programs. If you're using different editing software, check the manufacturer's documentation for keyboard controls.

As you can see, some controls are fairly standardized across many different video-editing applications. In fact, the Spacebar also controls the Play/Stop/Pause function in professional-grade video editors like Adobe Premiere and Apple Final Cut Pro.

Table 8-1 Keyboard Controls for Video Playback

Command	*Apple iMovie*	*Microsoft Windows Movie Maker*	*Pinnacle Studio*
Play/Stop	Spacebar	Spacebar	Spacebar or K
Fast forward	⌘+]	Ctrl+Alt+Right arrow	L
Rewind	⌘+[	Ctrl+Q	J
One frame forward	Right arrow	Alt+Right arrow	X
One frame back	Left arrow	Alt+Left arrow	Y

If you don't like using the keyboard to try to move to just the right frame, you may want to invest in a multimedia controller such as the SpaceShuttle A/V from Contour A/V Solutions (`www.contouravs.com`). This device connects to your computer's USB port and features a knob and dial that you can use to precisely control video playback. It's so much easier than using the mouse or keyboard, you'll wonder how you ever got by without one! (I describe multimedia controllers in greater detail in Chapter 18.)

Trimming out the unwanted parts

Professional Hollywood moviemakers typically shoot hundreds of hours of footage just to get enough acceptable material for a two-hour feature film. Because the pros shoot a lot of "waste" footage, don't feel so bad if every single frame of video you shot isn't totally perfect either. As you preview your clips, you'll no doubt find bits that you want to cut from the final movie.

Consider "Scene 2" from the Chapter 8 sample clips folder on the CD-ROM that accompanies this book. The subject scratches his lip for a few moments at the beginning of the clip, which would be fine, except that it kind of looks like he's picking his nose. We can't have *that* in the final movie. Besides, the clip is about 11 seconds long, and we really only need about five seconds or so.

The solution is to trim the clip down to just the portion you want. The easiest way to trim a clip is to split it into smaller parts before you place it in your movie project:

1. **Open the clip you want to trim by clicking it in the browser, and move the play head to the exact spot where you want to split the clip.**

   Use the playback controls under the preview window to move the play head. Table 8-1 lists some keyboard shortcuts to help you move the play head more precisely. If you're using the Scene 2 sample clip, place the play head about four seconds into the clip. If you're using Pinnacle Studio, the timecode will actually read about 0:00:16.20 because the timecode is for the entire sample file and not just the selected clip.

2. **In Pinnacle Studio, right-click the clip in the browser and choose Split Scene from the menu that appears. In Apple iMovie, choose Edit⇨Split Video Clip at Playhead.**

   You now have two clips where before you had only one.

3. **To split the second clip again, choose the second clip by clicking it in the browser and move the play head about five seconds forward in the clip.**

   Again, in Pinnacle Studio the timecode will actually read about 0:00:21.20 because the timecode displayed is for the entire sample file.

4. **Repeat Step 2 to split the clip again.**

You will now have three clips created from the one original clip (as shown in Figure 8-5). Splitting clips like I've shown here isn't the only way to edit out unwanted portions of video. You can also trim clips once they're placed in the timeline of your movie project (a process described later in this chapter). But splitting the clips before you add them to a project is often a much easier way to work because the unwanted parts are split off into separate clips that you can use (or not use) as you wish. In the next section, I show you how to add clips to the timeline or storyboard of your editing program to actually start turning your clips into a movie.

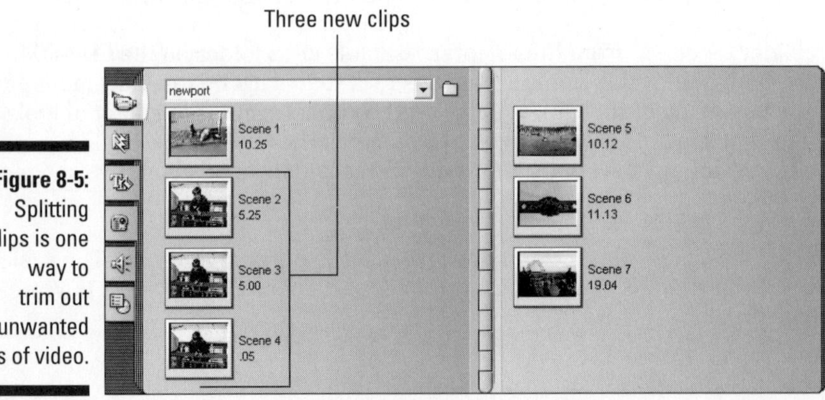

Figure 8-5: Splitting clips is one way to trim out unwanted bits of video.

Understanding timecode

A video image is actually a series of still frames that flash rapidly by on-screen. Every frame is uniquely identified with a number called a *timecode.* All stored locations and durations of all the edits you perform on a movie project use timecodes for reference points, so a basic understanding of timecode is important. You'll see and use timecode almost every time you work in a video-editing program like Pinnacle Studio or Apple iMovie. Timecode is often expressed like this:

```
hours : minutes : seconds :
    frames
```

The fourteenth frame of the third second of the twenty-eighth minute of the first hour of video is identified as:

```
01:28:03:13
```

You already know what hours, minutes, and seconds are. *Frames* aren't units of time measurement, but rather, the individual still images that make up your video. The frame portion of timecode starts with zero (00) and counts up to a number determined by the frame rate of the video. In PAL video, frames are counted from 00 to 24 because the frame rate of PAL is 25 frames per second (fps). In NTSC, frames are counted from 00 to 29. The NTSC and PAL video standards are described in greater detail in Chapter 3.

"Wait!" you exclaim. "Zero to 29 adds up to 30, not 29.97."

You're an observant one, aren't you? As mentioned in Chapter 3, the frame rate of NTSC video is 29.97 fps. NTSC timecode actually skips the frame codes 00 and 01 in the first second of every minute, except every tenth minute. Work it out (you may use a calculator), and you see that this system of reverse leap-frames adds up to 29.97 fps. This is called *drop-frame* timecode. In some video-editing systems, drop-frame timecode is expressed with semicolons (;) between the numbers instead of colons (:). Thus, in drop-frame timecode, the fourteenth frame of the third second of the twenty-eighth minute of the first hour of video is identified as

```
01;28;03;13
```

Why does NTSC video use drop-frame timecode? Back when everything was broadcast in black and white, NTSC video was an even 30 fps. For the conversion to color, more bandwidth was needed in the signal to broadcast color information. By dropping a couple of frames every minute, there was enough room left in the signal to broadcast color information, while at the same time keeping the video signals compatible with older black-and-white TVs.

Although the punctuation (for example, colons or semicolons) for separating the numerals of timecode into hours, minutes, seconds and frames is fairly standardized, some video-editing programs still go their own way. Pinnacle Studio, for example, uses a decimal point between seconds and frames. But whether the numbers are separated by colons, decimals, or magic crystals, the basic concept of timecode is the same.

Don't worry! Trimming a clip doesn't delete the unused portions from your hard drive. When you trim a clip, you're actually setting what the video pros call *in points* and *out points*. The software uses virtual markers to remember which portions of the video you chose to use during a particular edit. If you want to use the remaining video later, it's still on your hard drive, ready for use.

If you want, you can usually unsplit your clips that you have split as well. In Pinnacle Studio, hold down the Ctrl key and click on each of the clips that you split earlier. When each clip is selected, right-click the clips and choose Combine Scenes. Unfortunately, Apple iMovie doesn't have a simple tool for recombining clips that you have split. If you just split a clip, you can undo that action by choosing Edit⇨Undo or pressing ⌘+Z.

Turning Your Clips into a Movie

You're probably wondering when the fun begins. This is it! It's finally time to start assembling your various video clips into a movie. Most video-editing programs provide the same two basic tools to help you assemble a movie:

- **Storyboard:** This is where you throw clips together in a basic sequence from start to finish — think of it as a rough draft of your movie.
- **Timeline:** After your clips are assembled in the storyboard, you can switch over to the timeline to fine-tune the movie and make more advanced edits. The timeline is where you apply the final polish.

The storyboard and timeline are basically just two different ways of showing the same thing. In most editing programs — including iMovie, Studio, and Windows Movie Maker — you can toggle back and forth between the storyboard and timeline whenever you want. Some people prefer to use one or the other exclusively; for now, starting with the storyboard will keep it simple.

Visualizing your project with storyboards

If you've ever watched a "making of" documentary for a movie, you've probably seen filmmakers working with a storyboard. It looks like a giant comic strip where each panel illustrates a new scene in the movie. The storyboard in your video-editing program works the same way. You can toss scenes in the storyboard, move them around, remove scenes again, and just generally put your clips into the basic order in which you want them to appear in the movie. The storyboard is a great place to visualize the overall concept and flow of your movie.

To add clips to the storyboard, simply drag them from the clip browser down to the storyboard at the bottom of the screen. As you can see in Figure 8-6, I have added five clips to the iMovie storyboard.

The storyboard should show a series of thumbnails, as shown in Figure 8-6. If your screen doesn't quite look like this, you may need to switch to the storyboard view. (Figure 8-6 shows the buttons for toggling between storyboard view and timeline view in iMovie; the equivalent buttons for Studio are shown in Figure 8-7.)

Figure 8-6: The storyboard is a good place to start assembling your project.

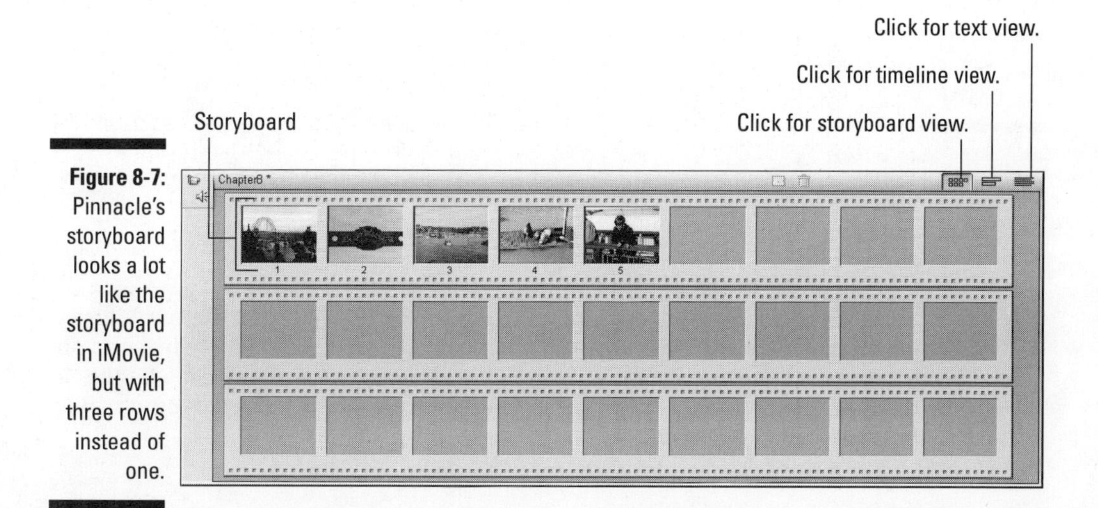

Figure 8-7: Pinnacle's storyboard looks a lot like the storyboard in iMovie, but with three rows instead of one.

The *storyboard* is possibly one of the most aptly named items in any video-editing program because the thumbnails actually do tell the basic story of your movie. The storyboard is pretty easy to manipulate. If you don't like the order of things, just click-and-drag clips to new locations. If you want to remove a clip from the storyboard in iMovie, Studio, or most any other editing program, click the offending clip once to select it and then press Delete on your keyboard.

Using the timeline

Some experienced editors prefer to skip the storyboard and go straight to the timeline because it provides more information and precise control over your movie project. To switch to the timeline, click the timeline button. (See Figure 8-6 for the location of the button in iMovie; see Figure 8-7 to find where the button is located in Studio.)

One of the first things you'll probably notice about the timeline is that not all clips are the same size. Consider Figure 8-8, which shows the timeline view of the same project shown in Figure 8-7. In the timeline view, the width of each clip represents the length (in time) of that clip, unlike in storyboard view, where each clip appears to be the same size. In timeline view, longer clips are wider, shorter clips are narrower.

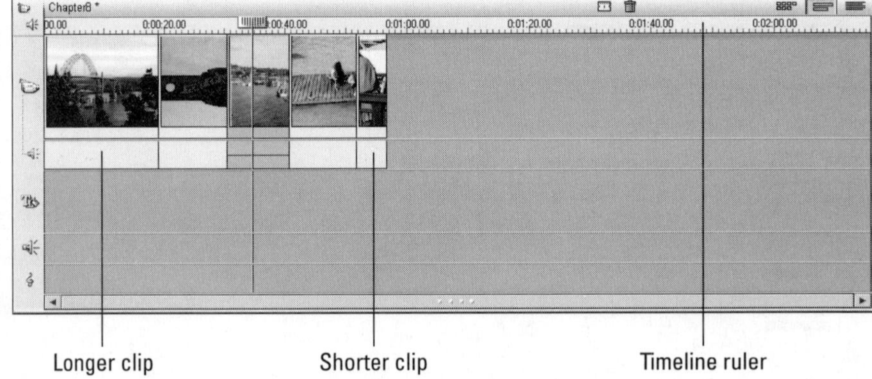

Figure 8-8: The timeline provides a bit more information about your movie project.

Adding clips to the timeline

Adding a clip to the timeline is a lot like placing clips in the storyboard. Just use drag-and-drop to move clips from the clip browser to the timeline. As you can see in Figure 8-9, I am inserting a clip between two existing clips on the timeline. Clips that fall after the insert are automatically shifted over to make room for the inserted clip.

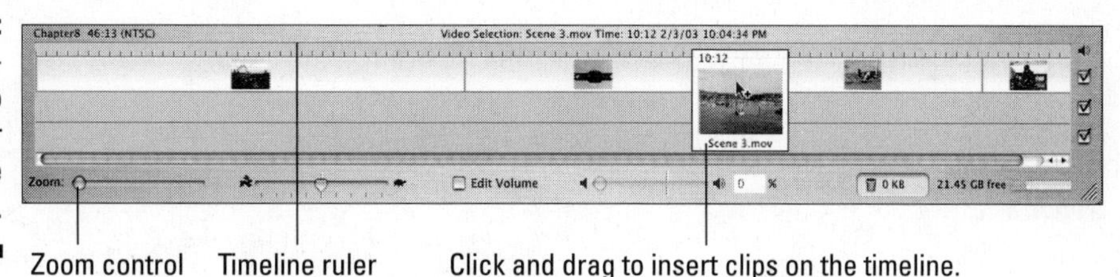

Figure 8-9: Use drag-and-drop to place your clips in the timeline.

Zooming in and out on the timeline

Depending on how big your movie project is, you may find that clips on the timeline are often either too wide or too narrow to work with effectively. To rectify this situation, adjust the zoom level of the timeline. You can either zoom in and see more detail, or zoom out and see more of the movie. To adjust zoom, follow these steps:

- **Apple iMovie:** Adjust the Zoom slider control in the lower left corner of the timeline (refer to Figure 8-9).
- **Microsoft Windows Movie Maker:** Click the Zoom In or Zoom Out magnifying glass buttons above the timeline, or press Page Down to zoom in and Page Up to zoom out.
- **Pinnacle Studio:** Press the + key to zoom in, or press the - key to zoom out. Alternatively, hover your mouse pointer over the timeline ruler so the pointer becomes a clock, and then click-and-drag left or right on the ruler to adjust zoom.

Tracking timeline tracks

As you look at the timeline in your editing software, you'll notice that it displays several different tracks. Each track represents a different element of the movie — video resides on the video track; audio resides on the audio track. You may have additional tracks available as well, such as title tracks (see Chapter 9 for more on adding titles) or music tracks (see Chapter 10 for coverage on how to add music to your movie). Figure 8-10 shows the tracks used in the Pinnacle Studio timeline.

Some advanced video-editing programs (such as Adobe Premiere and Final Cut Pro) allow you to have many separate video and audio tracks in a single project. This advanced capability is useful for layering many different elements and performing some advanced editing techniques (described in Chapters 11 and 17).

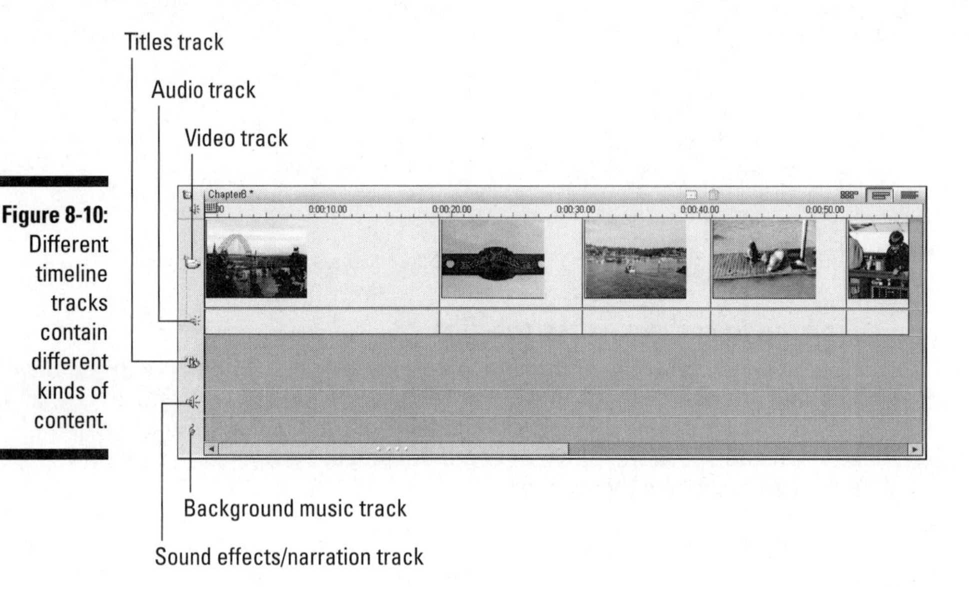

Figure 8-10: Different timeline tracks contain different kinds of content.

When you record and capture video, you usually capture audio along with it. All the Chapter 8 sample clips from the companion CD-ROM include audio tracks. When you place one of these video clips in the timeline (see Figure 8-10), the accompanying audio appears just underneath it in the audio track. Seeing the audio and video tracks separately is important for a variety of editing purposes. (Chapter 10 describes working with audio in greater detail.)

Locking timeline tracks in Pinnacle Studio

Pinnacle Studio offers a handy *locking* feature on timeline tracks. Locking the track doesn't prevent burglars from stealing it late at night, but it does allow you to temporarily protect a track from changes as you manipulate other tracks. For example, if you want to delete the audio track that came with some video, but you don't want to delete the video itself, follow these steps:

1. **Click the track header on the left side of the timeline.**

   A lock icon appears on the track header, and a striped gray background is applied to that track. In Figure 8-11, I have locked the main video track.

2. **Perform edits on other tracks.**

   For example, if you want to delete the audio track for one of your video clips, click the audio clip once to select it (as shown in Figure 8-11), and then press Delete on your keyboard. The audio portion of the clip disappears, but the video clip remains unaffected.

3. **Click the track header again to unlock the track.**

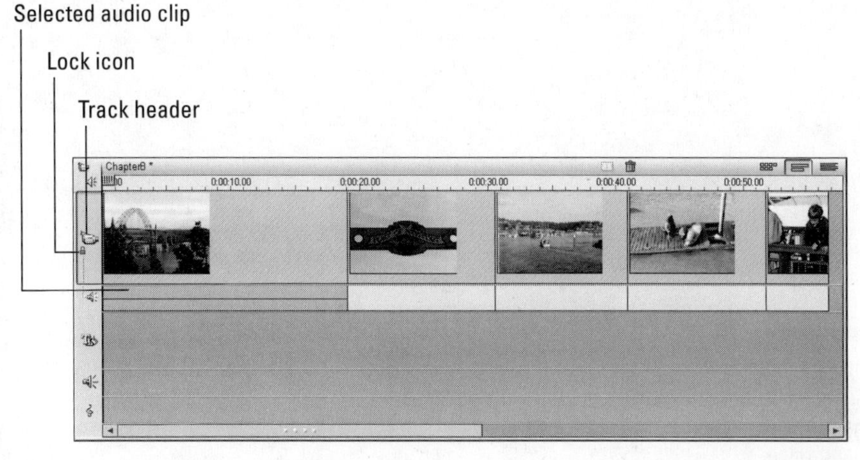

Figure 8-11: Locking some of your tracks will protect them from changes while you edit other tracks.

You can undo changes (such as deleting an audio clip) in Pinnacle Studio by pressing Ctrl+Z. If you followed the steps just given, press Ctrl+Z once to re-lock the track, and then press Ctrl+Z again to undelete the audio clip.

What did iMovie do with my audio?

Apple iMovie 3 offers some useful improvements over previous versions of the software — and a few changes that are less welcome. One thing I find a little aggravating is that the timeline does not automatically show the audio clips that accompany video clips. Take a look back at Figure 8-9 to see what I mean. Each clip in the timeline includes both audio and video, but the timeline shows only a single track.

To view combined audio and video clips separately in iMovie, you must extract the audio from each video clip individually. To do so, follow these steps:

1. **Click once on a clip in the timeline to select it.**
2. **Choose Advanced➪Extract Audio, or press ⌘+J.**

   The audio will now appear as a separate clip in the timeline, as shown in Figure 8-12.

3. **Repeat Steps 1 and 2 for each clip in the timeline.**

It may be a good idea to wait until later (like, when you're done editing the video portion of the movie) to extract audio from your video clips. If you still need to trim the video clip (as described later in this chapter), you'll have to trim the audio clip separately if it has been extracted.

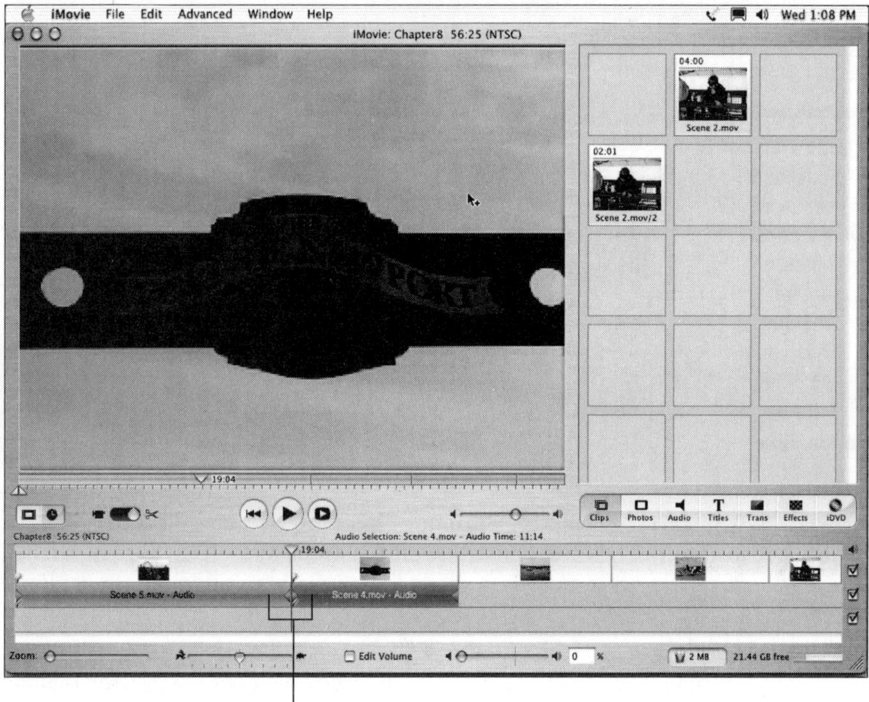

Figure 8-12: To work with audio and video separately in iMovie 3, extract audio clips from the video clips.

Fine-Tuning Your Movie in the Timeline

After you've plopped a few clips into your timeline or storyboard, you're ready to fine-tune your project. This fine-tuning is what turns your series of clips into a real movie. Most of the edits described in this section require you to work in the timeline, although if you want to simply move clips around without making any edits or changes, you'll probably find that easiest in the storyboard. This is especially true if you're using iMovie. To move a clip, simply click-and-drag it to a new location as shown in Figure 8-13.

Trimming clips in the timeline

Dropping clips into the storyboard or timeline is a great way to assemble the movie, but a lot of those clips probably contain some material that you don't want to use. Consider the bridge scene in the Chapter 8 sample clips from the CD-ROM. The clip is about 19 seconds long, which is much longer than we really want. As you play through the clip, you'll also notice that about 15 seconds into the clip, the camera moves. We don't want to include that in the movie, so let's trim this clip down to just the first 5 seconds.

Figure 8-13: Click-and-drag clips in the storyboard to move them to new locations in the movie.

To move a clip, simply drag and drop in storyboard view.

Trimming clips in Pinnacle Studio

The easiest way to trim clips in the Studio timeline is to use the Clip Properties window. To reveal this window, double-click a clip in the timeline. The properties window will appear above the timeline as shown in Figure 8-14. The left pane of the properties window shows the in point frame, which is the first frame of the clip. The right pane shows the out point, or the end of the clip. To adjust the in and out points, click and drag the in-and out-point razor tools back and forth. As you can see in Figure 8-14, I have dragged the out-point razor so only the first five seconds of the clip will play.

If you're using the sample clips from the CD-ROM, open the properties window for the bridge scene and adjust the out point as shown in Figure 8-14. You can also adjust the in and out points by typing new numbers in the time-code indicators under each pane.

The playback controls in the Clip Properties window include a Play Clip Continuously button. Click this button to preview the clip so it loops over and over continuously. This can help you better visualize the effects of any changes you make to the in and out points.

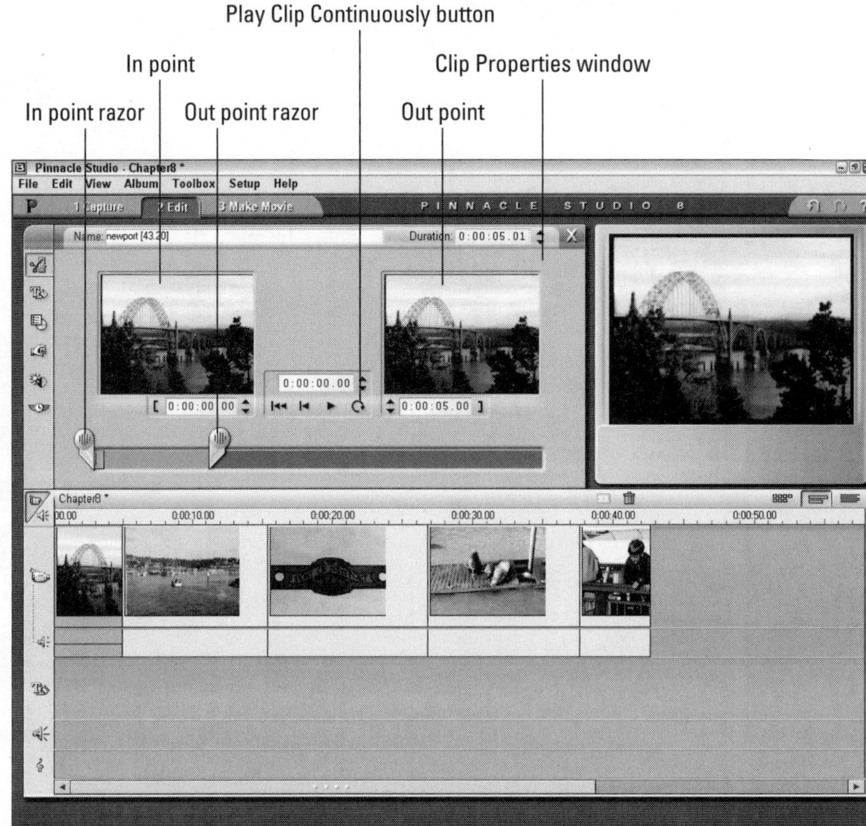

Figure 8-14: Fine-tune the in points and out points for clips in the timeline by using the Properties window.

When you're done trimming the clip, click the Close (X) button in the upper right corner of the Clip Properties window. The Clip Properties window closes and the length of your clip in the timeline will be changed accordingly.

Trimming clips in Apple iMovie

iMovie's approach to trimming is typical Apple — simple and effective. iMovie uses the preview window for clip-trimming operations. If you look closely at the lower-left corner of the preview window, you'll see two tiny little triangles, as shown in Figure 8-15. These are in-point and out-point markers; you can use them to trim clips in the timeline.

To trim a clip, follow these steps:

1. **Click a clip in the timeline to select it.**

   When you click the clip, it should load into the preview window. If it does not, make sure that the desired clip is actually selected in the timeline. The clip should turn blue when it is selected.

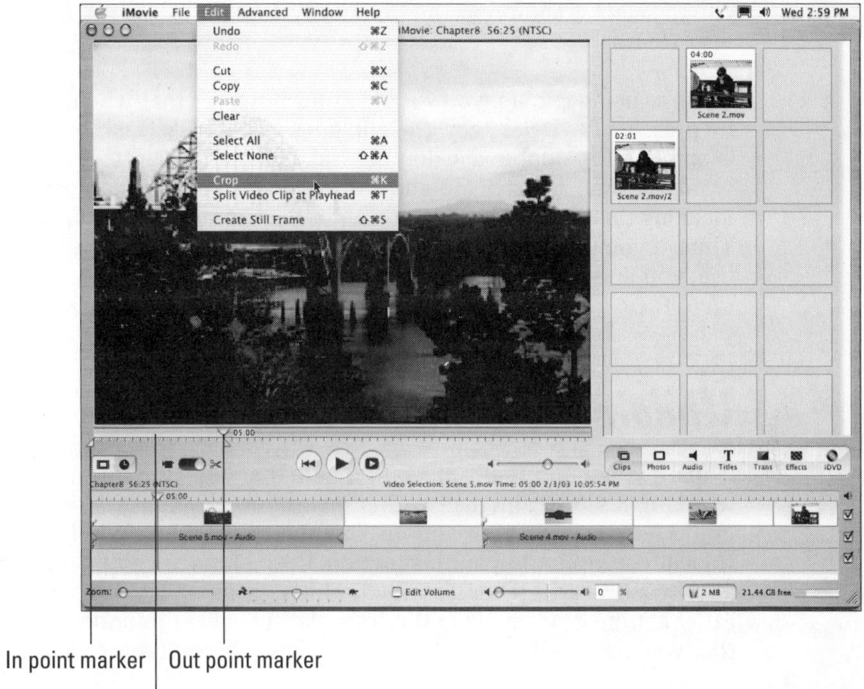

Figure 8-15: Use the iMovie preview window to trim clips in the timeline.

2. **Click-and-drag the out-point marker in the lower-left corner of the preview window to a new location along the playback ruler, as shown in Figure 8-15.**
3. **Click-and-drag the in-point marker to a new position if you wish.**

   The selected portion of the clip — that is, the portion between the in and out points — will turn yellow in the preview window playback ruler.
4. **Choose Edit⇨Crop as shown in Figure 8-15.**

   The clip will be trimmed (*cropped*) down to just the portion you selected with the in and out points. Subsequent clips in the timeline will automatically shift to fill in the empty space on the timeline.

If you trim a video clip from which you have extracted the audio, only the video clip will be trimmed; you'll have to trim the audio separately.

Removing clips from the timeline

Changing your mind about some clips that you placed in the timeline is virtually inevitable, and fortunately removing clips is easy. Just click the offending clip to select it and press the Delete key on your keyboard. Poof! The clip disappears.

If you're using iMovie, you should know however that when you delete a clip by pressing the Delete key, the clip goes to iMovie's trash bin. After the trash is emptied (by double-clicking the trash icon at the bottom of the iMovie program window), the deleted clip will no longer be stored on your hard drive, meaning that if you decide you want it back later, you'll have to re-capture it. Thus, if you simply want to remove a clip from the timeline, I recommend that you first switch to the storyboard, and then drag the unwanted clip back up to the clip browser so you can use it again later if you want.

Undoing what you've done

Oops! If you didn't really *mean* to delete that clip, don't despair. Just like word processors (and many other computer programs), video-editing programs let you undo your actions. Simply press Ctrl+Z (Windows) or ⌘+Z (Macintosh) to undo your last action. Both Pinnacle Studio and Apple iMovie allow you to undo several actions, which is helpful if you've done a couple of other actions since making the "mistake" you want to undo. To redo an action that you just undid, press Ctrl+Y (Windows) or Shift+⌘+Z (Macintosh).

Another quick way to restore a clip in iMovie to its original state — regardless of how long ago you changed it — is to select the clip in the timeline and then choose Advanced⇨Restore Clip. The clip reverts to its original state.

Adjusting playback speed

One of the coolest yet most unappreciated capabilities of video-editing programs is the ability to change the speed of video clips. Changing the speed of a clip serves many useful purposes:

- Add drama to a scene by slowing down the speed to create a "slow-mo" effect.
- Make a scene appear fast-paced and action-oriented (or humorous, depending upon the subject matter) by speeding up the video.
- Help a given video clip better fit into a specific time frame by speeding it up or slowing it down slightly. For example, you may be trying to time a video clip to match beats in a musical soundtrack. Sometimes this can be achieved by slightly adjusting the playback speed of the video clips.

Of course, you want to carefully preview any speed changes you make to a video clip. Depending on the software, you could encounter some jittery video images or other problems when you play around with speed adjustments.

Pay special attention to audio clips when you adjust playback speed. Even though a small speed adjustment might be barely perceptible in a video clip, even the tiniest speed changes have radical effects on the way audio sounds. Usually, when I adjust video speed, I discard the audio portion of that clip. Pinnacle Studio is unusual in that when you change the speed of a video clip, the audio portion of that clip is automatically discarded.

Adjusting playback speed in Apple iMovie

Changing playback speed in iMovie couldn't be easier. If you look closely, you'll see a slider adjustment for playback speed right on the timeline. To adjust speed, follow these steps:

1. **Switch to the timeline (if you aren't there already) by clicking the timeline view button (refer to Figure 8-6).**
2. **Click the clip that you want to adjust to select it.**

   The clip should turn blue when it is selected.
3. **Adjust the speed slider at the bottom of the timeline, as shown in Figure 8-16.**

Figure 8-16: Use the speed slider at the bottom of the timeline to adjust playback speed.

To speed up the clip, move the slider toward the hare. Move the slider toward the tortoise to (surprise) slow down the clip.

4. **Play the clip to preview your changes.**
5. **Giggle at the way the audio sounds after your changes.**

If you don't want to include the audio portion of the clip after you've made speed changes, choose Advanced➪Extract Audio to extract the audio, and then delete the audio clip after it is extracted.

Another neat thing you can do to video clips in iMovie is reverse the playback direction. To do so, select the clip and choose Advanced➪Reverse Clip Direction. The clip will now play backward in your movie. To reverse it back to normal, just choose Advanced➪Reverse Clip Direction again.

Adjusting playback speed in Pinnacle Studio

Pinnacle Studio gives you pretty fine control over playback speed. You can also adjust Strobe if you want to create a stop-motion effect that you may or may not find useful. The only way to really know is to experiment, which you can do by following these steps:

1. **Switch to the timeline (if you're not there already) by clicking the timeline view button (refer to Figure 8-7).**
2. **Double-click a clip in the timeline to open the Clip Properties window.**
3. **Click the Vary Playback Speed tool on the left side of the Clip Properties window (as in Figure 8-17).**

   The Vary Playback Speed controls appear as shown in Figure 8-17.

4. **Move the Speed slider left to slow down the clip, or move it to the right to speed up the clip.**

   An adjustment factor appears above the slider. Normal speed is shown as 1.0 X. Double speed would be 2.0 X, and so on. If you slow the playback speed down, a fraction will appear instead. For example, half speed will be indicated as 5/10 X.

5. **Move the Strobe slider to add some strobe effect.**
6. **Click Play in the preview window to view your changes.**
7. **If you don't like your changes and want to revert to the original speed or strobe setting, click one of the Reset buttons.**
8. **Click the Close (X) button in the upper-right corner of the Clip Properties window when you're done making changes.**

The Clip Properties window closes. If you speeded up playback of the clip, the clip will now appear narrower in the timeline. If you slowed down playback speed, the clip will be wider in the timeline.

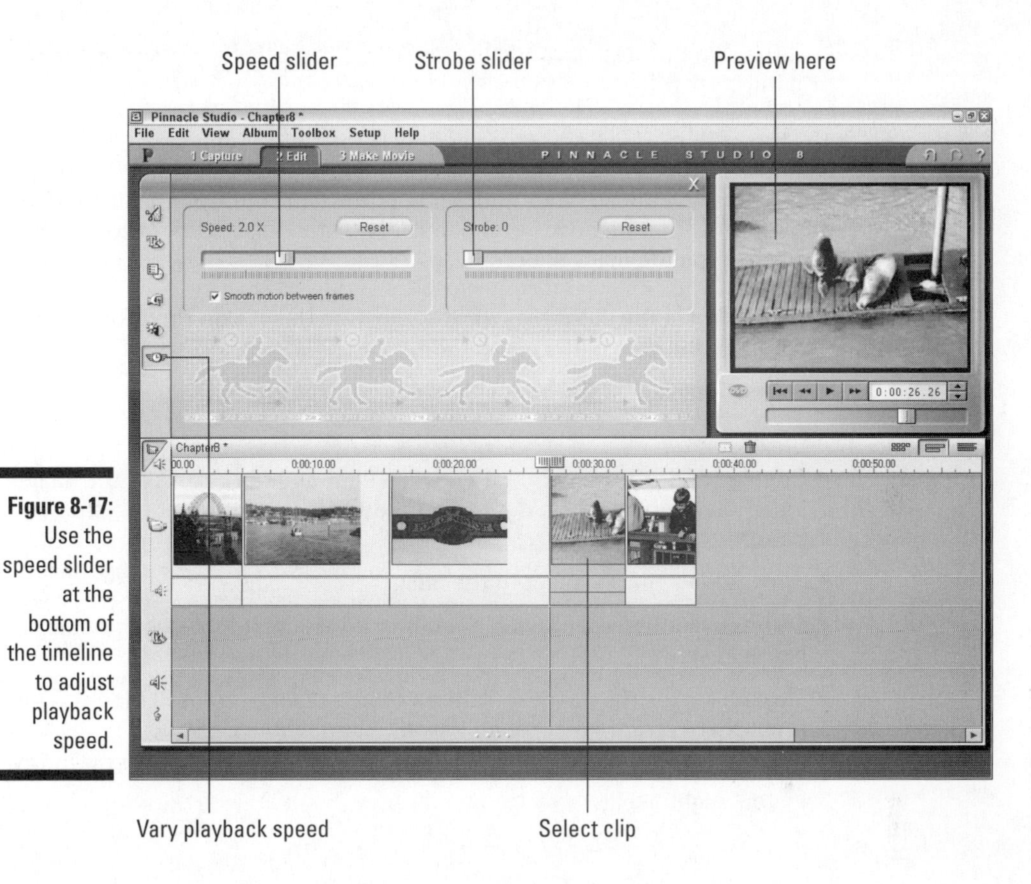

Figure 8-17: Use the speed slider at the bottom of the timeline to adjust playback speed.

Fixing Color and Light Issues

Even if you follow all the best advice for shooting great video, you will probably wind up with some video that has ugly coloration or poor lighting. Fortunately, modern editing programs give you some tools to correct some image quality issues.

Adjusting image qualities in Pinnacle Studio

Pinnacle Studio provides a pretty good selection of image controls. You can use these controls to improve brightness and contrast, adjust colors, or add a stylized appearance to a video image. To access the color controls, double-click a clip that you want to improve or modify. When the Clip Properties dialog box appears, click the Adjust Color or Add Visual Effects tab on the left side of the Clip Properties window. The Color Properties window appears as shown in Figure 8-18.

At the top of the Color Properties window is the Color Type menu. Here you can choose a basic color mode for the clip. The normal mode is All Colors, but you can also choose Black and White (shown in Figure 8-18), Single Hue, or Sepia. Below that menu are eight controls (listed in the order they appear) that you manipulate using sliders:

- **Hue:** Adjusts the color bias for the clip. Use this if skin tones or other colors in the image don't look right.
- **Saturation:** Adjust the intensity of color in the image.
- **Brightness:** Makes the image brighter or darker.
- **Contrast:** Adjust the contrast between light and dark parts of the image. If the image appears too dark, use brightness and contrast together to improve the way it looks.
- **Blur:** Adds a blurry effect to the image.
- **Emboss:** Simulates a carving or embossed effect. It looks cool, but is of limited use.
- **Mosaic:** Makes the image look like a bunch of large colored blocks. This is another control you probably won't use a whole lot.
- **Posterize:** Reduces the number of different colors in an image. Play with it; you might like it.

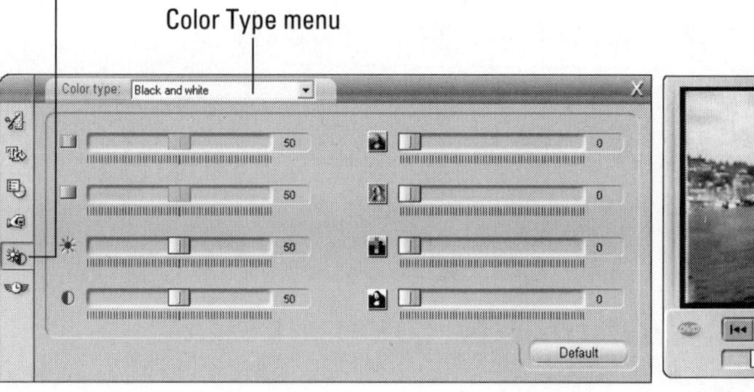

Figure 8-18: Although this book is printed without color anyway, trust me: I converted this clip to black and white.

Most of your clips are probably fine just the way they are, but it's good to know that these controls are there if you need them. And if, after making a bunch of changes, you decide that you liked the clip better the way it was, just click the Default button to return the clip back to its original settings.

Modifying light and color in Apple iMovie

Although Apple iMovie doesn't have specific image controls (as does, say, Studio), you can still modify color and light characteristics using some of iMovie's effects. Start by selecting a clip that you want to adjust, and then click the Effects button in the upper-right portion of the timeline. The Effects window appears (as shown in Figure 8-19). Then you can use any of several effects to improve the appearance of the clip. Most effects have controls you can adjust by moving sliders. You can also control how the effect starts and finishes. (We get into controlling effects timing more in Chapter 11.) Effects that can modify color and lighting include the following:

- **Adjust Colors:** Allows you to adjust hue, saturation (color), and lightness.
- **Aged Film:** If the clip looks really bad, you can avoid blame by applying this effect to make it look like it's from really old film. ("See, it's not my fault that the colors are all wrong — this was shot on 8mm film 40 years ago!") Your secret is safe with me. The Aged Film effect lets you adjust three different factors:
  - The Exposure slider lets you make the aged effect appear lighter or darker.
  - The Jitter slider controls how much the video image "jitters" up and down. Jitter makes the clip look like film that is not passing smoothly through a projector.
  - The Scratches filter lets you adjust how many film "scratches" appear on the video image.
- **Black & White:** Converts the clip to a black and white image.
- **Brightness & Contrast:** Adjusts brightness and contrast in the image. (In Figure 8-19, I have increased brightness and contrast to improve the appearance of a backlit video clip.) Separate sliders let you control brightness and contrast separately.
- **Sepia Tone:** Gives the clip an old-fashioned sepia look.
- **Sharpen:** Sharpens an otherwise blurry image. A slider control lets you fine-tune the level of sharpness that is applied.
- **Soft Focus:** Gives the image a softer appearance, simulating the effect of a soft light filter on the camera. Three slider controls let you customize the Soft Focus effect:

- The Softness slider controls the level of softness. Move the Softness control towards the Lots end of the slider for a dream-sequence look.
- The Amount slider controls the overall level of the Soft Focus effect.
- The Glow slider increases or decreases the soft glow of the effect. Setting the Glow slider towards High tends to wash out the entire video clip.

To see a full-screen preview of an effect, click Preview. If you are happy with the effect, click Apply to apply the effect to the clip. When you click Apply, you may see a red progress bar appear on the clip. This shows the progress of the *rendering,* the process that actually applies the effect to the clip. The rendering process actually creates a temporary file on your hard disk that iMovie uses to show how the clip looks after the effect has been applied.

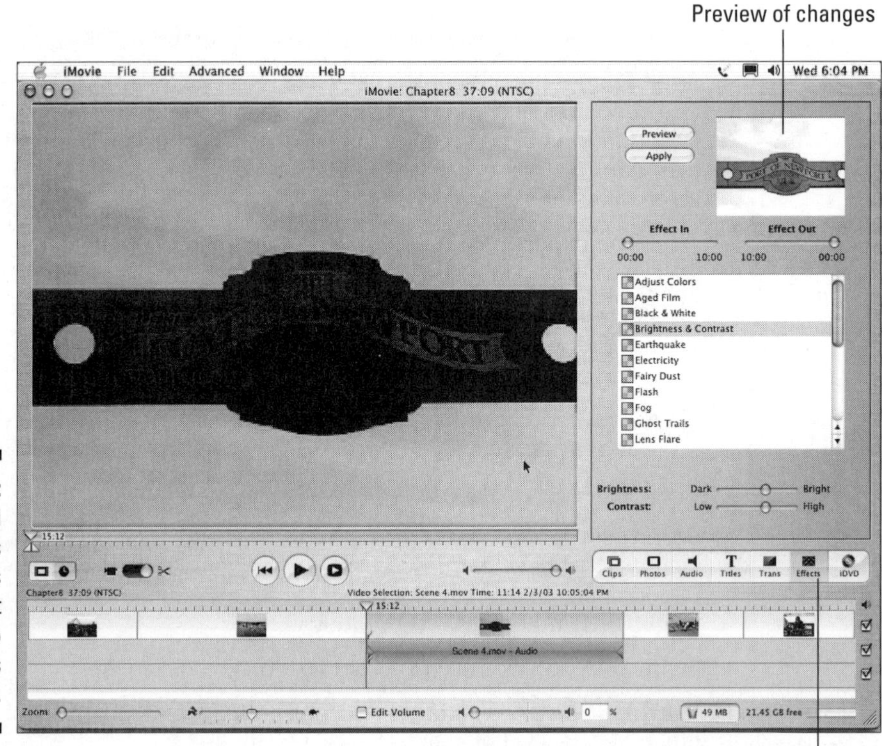

Figure 8-19: I used iMovie's Brightness & Contrast effect to improve this backlit clip.

Chapter 9

Using Transitions and Titles

In This Chapter

- Using fades and transitions between video clips
- Using predefined title styles in Pinnacle Studio
- Creating your own title styles in Studio
- Working with titles in iMovie

Anyone with two VCRs wired together can dub from tape to tape, copying only the desirable scenes, and call it *editing*. But the results will be pretty sloppy, especially compared to the high-quality movies you can make using your computer. Modern video-editing programs like Apple iMovie and Pinnacle Studio give you powerful tools to help you add special elements to your projects. For example, you can add a transition that allows one clip to gracefully fade into the next, and you can add your own on-screen credits and notations using titles.

This chapter shows you how to use transitions and titles in your movie projects. These are two of the most popular tools offered by video-editing programs; unlike some other (more obscure) effects and tools, transitions and titles are things you'll probably actually use in every single movie.

If you don't have your own video clips to work with, you can still follow along with the steps in this chapter by using the Chapter 9 sample clips on the CD-ROM that accompanies this book. Import the Chapter 9 sample clips (three files for Mac users, one file for Windows) from the Chapter 9 folder. See Appendix A for instructions on finding and importing the sample clips.

Using Fades and Transitions Between Clips

One of the trickiest aspects of movie editing (for me, anyway) is making clean transitions between clips. Often the best transition is a simple, straight cut from one clip to the next. Other times, you want to fade gently from one scene to the next. Or you may want a more fancy transition — say, one that makes it look like the outgoing scene is being rolled apart like drapes to reveal the incoming scene behind it. Most transitions can be generally divided into a few basic categories:

- **Straight cut:** This is actually no transition at all. One clips ends and the next begins, poof! Just like that.
- **Fade:** The outgoing clip fades out as the incoming clip fades in. Fades are also sometimes called *dissolves*.
- **Wipe:** The incoming clip wipes over the outgoing clip using one of many possible patterns. Alternatively, the outgoing clip may wipe away to reveal the incoming clip.
- **Push:** The outgoing clip is pushed off the screen by the incoming clip.
- **3-D:** Some more advanced editing programs provide transitions that seem to work three dimensionally. For example, the outgoing clip might wrap itself up into a 3-D ball, which then spins and rolls off the screen. Pinnacle's Hollywood FX plug-ins for Studio provide many interesting 3-D transitions. See Appendix D for more on Studio plug-ins.

Whatever style of transition you want to use, modern video-editing programs like Apple iMovie and Pinnacle Studio make the process easy. But before you can use any transitions, you need a project that already has several clips in its timeline. If you're working with the Chapter 9 sample clips from the companion CD-ROM, place all three clips in the Timeline in order (Scene 1, Scene 2, and finally Scene 3). If you don't yet feel comfortable with editing clips into the timeline, check out Chapter 8. The following sections show you how to select and use transitions in your movie projects.

Choosing the best transition

When Windows Movie Maker first came out in 2000, choosing what type of transition to use between clips was easy because you only had two choices. You could either use a straight-cut transition (which is actually no transition at all) or a cross-fade/dissolve transition. If you wanted to use anything fancier, you were out of luck.

Thankfully, most modern video-editing programs — including Windows Movie Maker 2 — provide you with a pretty generous assortment of transitions. Transitions are usually organized in their own window or palette. Transition windows usually vary slightly from program to program, but the basics are the same.

How do you decide which transition is the best? The fancy transitions may look really cool, but I recommend restraint when choosing them. Remember that the focus of your movie project is the actual video content, not showing off your editing skills or the capabilities of your editing software. More often than not, the best transition is a simple dissolve. If you do use a fancier transition, I recommend using the same or a similar transition throughout your project. This will make the transitions seem to fit more seamlessly into the movie.

Reviewing iMovie's transitions

Apple iMovie offers a selection of 13 transitions from which to choose. That may not sound like a big number, but I think you'll find that iMovie's 13 transitions cover the styles you're most likely to use anyway. To view the transitions that are available in iMovie, click the Trans button above the Timeline. A list of transitions appears, as shown in Figure 9-1.

As you look at the list of transitions, most of the names probably look pretty foreign to you. Names like "Circle Closing," "Radial," and "Warp Out" are descriptive, but really the only way to know how each transition will look is to preview it. To do so, click the name of a transition in the list. A small preview of the transition briefly appears in the transition preview window.

Figure 9-1: iMovie 3 provides a good selection of transitions that you can use in your projects.

Oops! You missed it. Click the transition's name again. Wow, it sure flashes by quickly, doesn't it? If you'd like to see a larger preview, click the Preview button. A full-size preview appears in iMovie's main viewer screen.

If your transition preview window shows nothing but a black screen when you click a transition, move the mouse pointer down and click a clip in the timeline or storyboard to select it. The selected clip should now appear when you preview a transition.

Previewing transitions in Pinnacle Studio

Pinnacle Studio comes with 142 (yes, 142) transitions. In fact, so many transitions are provided with Studio that they don't all fit on one page. To see a list of Studio's transitions, click the Show Transitions tab on the left side of the album, as shown in Figure 9-2. There are several pages of transitions, and you can view additional pages by clicking the arrows in the upper-right corner of the album.

To preview a transition, simply click it in the album window. A preview of the transition will appear in the viewer window to the right of the album. A blue screen labeled "A" represents the outgoing clip, whereas the incoming clip is represented by the orange "B" clip. In Figure 9-2, I am previewing a wipe that uses a heart pattern.

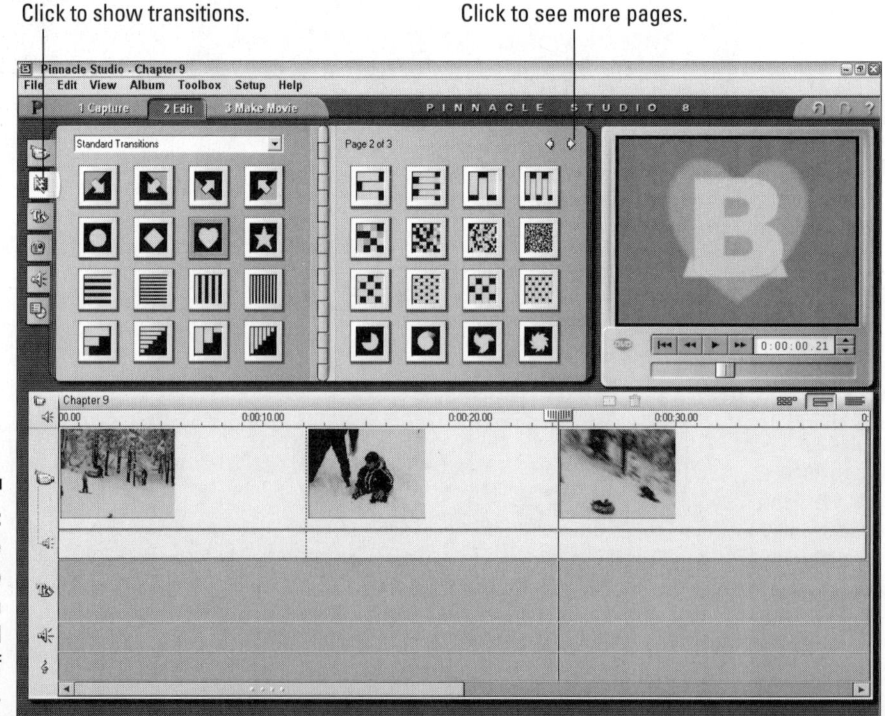

Figure 9-2: Pinnacle Studio comes with several pages of transitions.

Adding a transition to the timeline

Adding a transition to a project is pretty easy and works the same way in almost every video-editing program available. For now, add a simple dissolve (also called a fade) transition to a project. If your editing program currently shows the storyboard for your project, switch to the timeline (see Chapter 8 for more on the timeline and basic editing). Next, click-and-drag the Dissolve (in Pinnacle Studio) or Cross Dissolve (in iMovie) transition from the list of transitions and drop it between two clips on the timeline. If you're working with the Chapter 9 sample clips from the CD-ROM, drop the transition between the first two scenes. The transition will now appear in the timeline between the clips.

The appearance of the transition will vary slightly depending on the editing program you are using. As you can see in Figure 9-3, Pinnacle Studio displays the transition as a clip in the Timeline. Apple iMovie, on the other hand, uses a special transition icon that overlaps the adjacent clips as shown in Figure 9-4.

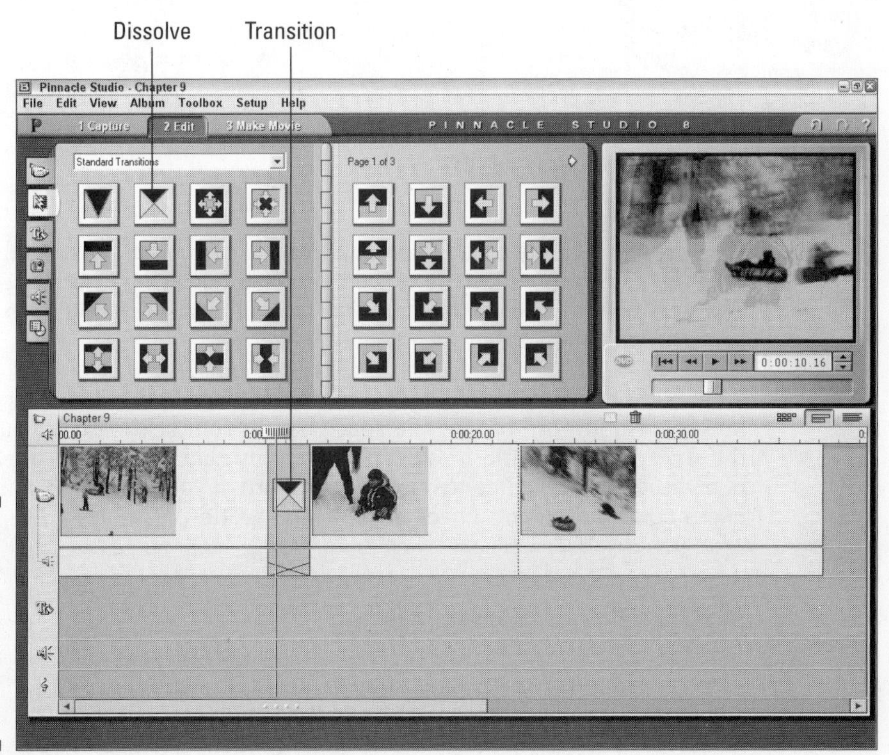

Figure 9-3: Transitions look like clips in the Pinnacle Studio timeline.

Figure 9-4: The iMovie transition icon.

iMovie uses a special transition icon.

When you first apply a transition in iMovie, the program must render the transition before it can be viewed. I explain rendering in more detail in Chapter 13, but basically it's a process that allows the computer to play back the transition at full speed and quality.

To preview the transition, simply play the timeline by clicking Play under the preview window or pressing the space bar on your keyboard. If you don't like the style of the transition, you can delete it by clicking the transition to select it, and then pressing Delete on your keyboard. If you think the transition just needs some fine-tuning, check out the next section.

Adjusting transitions

Video transitions usually have some features or attributes that can be adjusted. Most important, perhaps, is the length of the transition. The default length for most transitions is about two seconds. If you added a dissolve transition between the Scene 1 and Scene 2 sample clips (as described in the previous section), the two-second interval covers the time when wisps of Scene 2 just barely start to appear, ending when the last trace of Scene 1 fades out of existence.

Sometimes a two-second transition is too long — or not long enough. When applied to the sample clips Scene 1 and Scene 2, a two-second dissolve really *is* too long because it obscures a spectacular crash at the end of Scene 1. The next two sections show you how to adjust the length of transitions and make other changes where possible.

Modifying iMovie transitions

The iMovie transition window doesn't just provide a place to store transitions; it also allows you to control them. To adjust the length of a transition, follow these steps:

1. **Click the transition in the timeline to select it.**
2. **Adjust the Speed slider in the transitions window.**

   If you're working with the Chapter 9 sample clips, adjust the transition between Scene 1 and Scene 2 down so it's only 15 frames long.

   The length of the transition will be shown in the preview window, as shown in Figure 9-5.

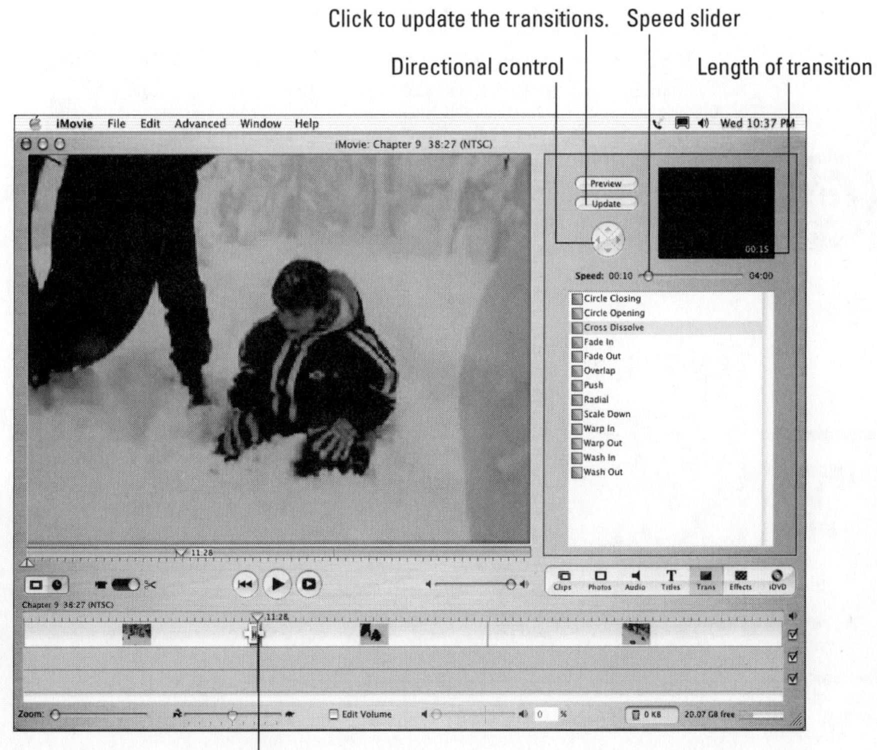

Figure 9-5: Use the Speed slider to adjust the length of the clip.

3. **Click Update in the transitions window to update the length of the transition in the timeline.**
4. **Click the Play button under the preview window to preview your results.**

Some iMovie transitions, such as Push, allow you to control the direction of travel for the transition. When you click the Push transition to select it in the transition window, the directional control (Figure 9-5) becomes active. Then you simply click the arrows in the directional control to change the direction in which the Push transition moves.

Adjusting transitions in Studio

Adjusting transitions in Pinnacle Studio works a lot like adjusting regular video clips. To modify a transition, here's the drill:

1. **Double-click the transition in the timeline.**

   The Clip Properties window appears above the timeline.

2. **Adjust the Duration of the transition in the upper-right corner of the window by clicking the up or down arrows next to the Duration field at the top of the Clip Properties window.**
3. **If you're working with the Chapter 9 sample clips, change the duration to 15 frames, as shown in Figure 9-6.**

   You can also reverse the direction of some transitions (for example, the Horizontal Snake Wipes on page two of the Standard Transitions) by placing a check in the Reverse check box. (If this box is grayed out, the current transition cannot be reversed.)

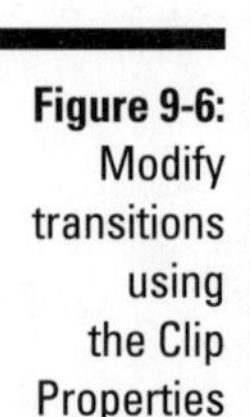

Figure 9-6: Modify transitions using the Clip Properties window.

Giving Credit with Titles

In their rush to get to the pictures, folks who are new to video editing often overlook the importance of titles. Titles — the words that appear on-screen during a movie — are critically important in many different kinds of projects. Titles tell your audience the name of your movie, who made it, who starred in it, who paid for it, and who baked cookies for the cast. Titles can also clue the audience in to vital details — where the story takes place, what time it is, even what year it is — with minimum fuss. And of course, titles can reveal what the characters are saying if they're speaking a different language.

Virtually all video-editing programs include tools to help you create titles. The following sections show you how to create and use effective titles in your movie projects.

The steps in the next few sections show you how to add titles to any movie project. If you don't have a project if your own to work on, you can use the Chapter 9 sample clips from the CD-ROM that accompanies this book. If you didn't use the clips in the previous section on transitions, import the Chapter 9 sample clips (Windows users will see only one file to import) into your video-editing program. Then place the three sample Scenes into the timeline in order.

Creating titles for your movies

It's easy to think of titles as just words on the screen. But think of the effects, both forceful and subtle, that well-designed titles can have. Consider the *Star Wars* movies, which all begin with a black screen and the phrase, "A long time ago, in a galaxy far, far away . . ." The simple title screen quickly and effectively sets the tone and tells the audience that the story is beginning. And then, of course, you get those scrolling words that float three-dimensionally off into space, immediately after that first title screen. A story floating through space is far more interesting than white text scrolling from the bottom to top of the screen, don't you think?

Titles can generally be put into two basic categories: Full-screen and overlays. Full-screen titles like the one shown in Figure 9-7 are most often used at the beginning and end of a movie, where the title appears over a background image or a solid color (such as a plain black screen). The full-screen title functions as its own element in the movie, as does any video clip. An overlay title makes words appear right over a video image, as shown in Figure 9-8.

Figure 9-7: A full-screen title stands alone and is often used at the beginning or end of a movie.

Figure 9-8: An overlay title appears right on a video image.

Making effective titles

When using titles in your movies, you should follow some basic guidelines to make them more effective. After all, funny or informative titles don't do much good if your audience can't read them. Follow these general rules when creating titles for your movies:

- **Less is more.** Try to keep your titles as brief and simple as possible.
- **White on black looks best.** When you read words on paper, black letters on white paper are easier to read. This rule does not carry over to video, however. In video images, light characters over a dark background are usually easier to read. An exception would be if you already have a relatively light background. In Figure 9-8, the darker colored title works well over the snow-covered background. But what if I want to put that same title near the top of the screen instead? The dark-colored title won't

work as well at the top of the screen because trees in the background will make it harder to read. As an alternative, I can create a small, dark background shape behind lighter colored text, as shown in Figure 9-9. (I'll show you how to control title colors and styles in the next couple of sections.)

Video displays such as TVs and computer monitors generate images using light. (You can think of a TV screen as a big light bulb.) Because TV displays emit light, full-screen titles almost never have black text on a white screen. Most people find staring at a mostly white TV screen about as unpleasant as staring at a lit light bulb: If you watch viewers look at a white TV screen, you may even see them squint. Squinting is bad, so stick to the convention of light words over a dark background whenever possible.

- **Avoid very thin lines.** In Chapter 3, I described the difference between interlaced and non-interlaced video displays. In an interlaced display — such as a TV — every other horizontal resolution line of the video image is drawn in a separate pass or field. If the video image includes very thin lines — especially lines that are only one pixel thick — interlacing could cause the lines to flicker noticeably, giving your viewers a migraine headache in short order. To avoid this, choose fonts that have thicker lines. For smaller characters, avoid serif fonts such as Times New Roman. *Serifs* (the extra strokes at the ends of characters in some fonts, such as the one you're reading right now) often have very thin lines that flicker on an interlaced video display.
- **Think big.** Remember that the resolution on your computer screen is a lot finer than what you get on a typical TV screen. Also, TV viewers typically watch video from longer distances than do computer users — say, across the room while plopped on the sofa, compared to sitting just a few inches away in an office chair. This means the words that look pleasant and readable on your computer monitor may be tiny and unintelligible if your movie is output to tape for TV viewing. Never be afraid to increase the size of your titles.

Figure 9-9: Small dark blobs or shapes behind titles can make them easier to read.

- **Think safety.** Most TVs suffer from a malady called *overscan*, which is what happens when some of the video image is cut off at the edges of the screen. You can think of TVs as being kind of like most computer printers. Most TVs can't display the far edges of the video image, just as most printers can't print all the way to the very edges of the paper. When you're working on a word processing document, your page has margins to account for the shortcomings of printers as well as to make the page more readable. The same applies to video images, which have *title safe margins*. To ensure that your titles show up on-screen and are not cut off at the edges, make sure your titles remain within this margin (also sometimes called the *title safe boundary*).

  Apple iMovie doesn't show these margins on-screen; instead, it automatically keeps your titles inside the margins. If you know your movie will be viewed only on computer screens, place a check mark next to the QT Margins option in the Titles window. This places the margins closer to the edge of the screen.

  Pinnacle Studio's Title Editor works like most video-editing programs: The margins are displayed on-screen and you have to remember to keep your titles inside them. The margins appear in the Title Editor window as red dotted lines, as shown in Figure 9-10.

Figure 9-10: Don't let your titles fall outside the title safe margins.

- **Play, play, play.** I cannot stress enough the importance of previewing your titles, especially overlays. Play your timeline after adding titles to see how they look. Make sure the title is readable, positioned well on-screen, and visible long enough to be read. If possible, preview the titles on an external video monitor (see Chapter 13 for more on using external monitors).

You can use transitions between full-screen titles and other video clips. Dissolves (described earlier in this chapter) often look nice when used to transition from a full-screen title to the video.

Using Studio's Title Editor

Pinnacle Studio comes with a remarkably advanced title designer considering the price. Studio's Title Editor is based on Pinnacle's Title Deko, a high-quality title designer used by many professional video editors. Studio also provides a selection of predesigned titles that are ready to drop right into any video project. To access these titles, click the Show Titles tab on the left side of the album. A selection of titles appears as shown in Figure 9-11. Click the arrow in the upper-right corner of the album to view additional pages.

To see a preview of the title, click it once in the album. The title will appear in the viewer window. To add a title to your project, simply drag it from the album down to the timeline just as if it was a video clip. In Figure 9-11,you can see that I have dropped a "Snow Boarding" title at the beginning of my timeline's video track. If the title is dropped on the video track, the title will be a full-screen title which by default has a plain black background.

If you use a predesigned title, chances are you'll want to change some of the text. For example, even though I like the style of the "Snow Boarding" title, I don't want the words *Snow Boarding* to appear because that's not what is happening in the video. To edit the title, double-click it after you have added it to the timeline. The Studio Title Editor will open. You can also open the Title Editor and create a new title by double-clicking any blank space in the Title track (see Figure 9-11) of the timeline. The next section shows you how to edit a title.

Editing titles

The Title Designer window is one of the most complex windows in the Studio software. If you are working with a predesigned title, double-click it in the timeline to change the text. The title will open in the title editor. To change the text, click once on the title and start typing. Highlight unwanted words and press the Delete key to delete them.

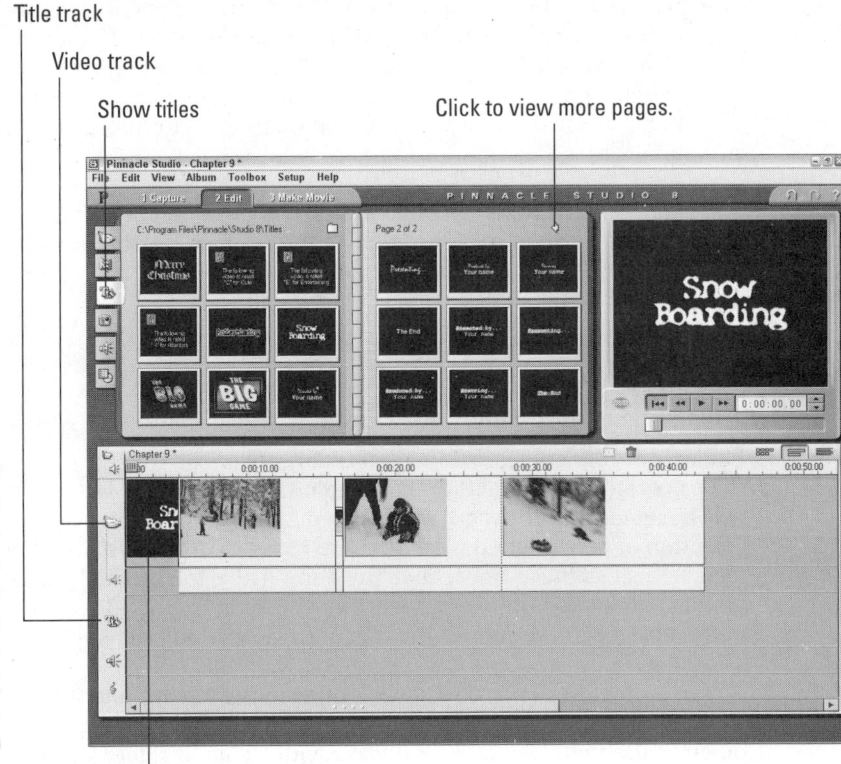

Figure 9-11: Predesigned titles make it easy to place titles quickly in your movies.

You can change the font style or size for the title if you don't like it. Select the text you want to change, and then choose a new font in the font menu at the top of the Title Editor. In Figure 9-12, I have changed the font to Trendy because I didn't like the Pretext font that was used in the "Snow Boarding" title. Figure 9-12 also shows the basic controls of the Title Designer. Some of the most important controls include

- **Roll:** Click this to make the title roll vertically up the screen. This is especially useful if you have a long list of credits that you want to roll at the end of the movie. If you choose the Roll option, the Title Editor allows you to scroll down the editing area to add more rows of titles. You can make a roll title as long as you want. Each time you add a new line of text to the bottom of a rolling title, the title screen gets a little longer.
- **Crawl:** Click this to make the title crawl across the screen from right to left. Crawl useful text across the bottom of the screen, such as, "Order now! Quantities are limited!" Like rolling titles, you can make a crawling title wider by simply adding more text to the right side of the title.

- **Text styling:** Select some text in your title, and then click one of these buttons to style the text. Just like in a word processor or almost any other computer program, click B to make the text bold, click I to make it italic, and click U to underline the text.
- **Text justification:** Click this button to open a submenu of text justification and alignment options. You can choose to align text left, right, or centered. If you click the Shrink to Fit button, the text will automatically shrink to fit in the text box if you make the box smaller. If you click Scale to Fit, the text shrinks or grows to fill the entire text box if you make it smaller or bigger. If you want long lines of text to automatically wrap to the next line, click the Word Wrap On button. Otherwise, click the Word Wrap Off button. Click the Close (X) button in the upper right corner of the Text justification menu to close the menu.
- **Add Text Field:** Click this button and then click-and-drag a text box in the editing area to create a new line of text from scratch.
- **Text styles:** Use these predefined styles to quickly format text. To apply a style to some existing text, select the text in the editing area and then click a style in the list.

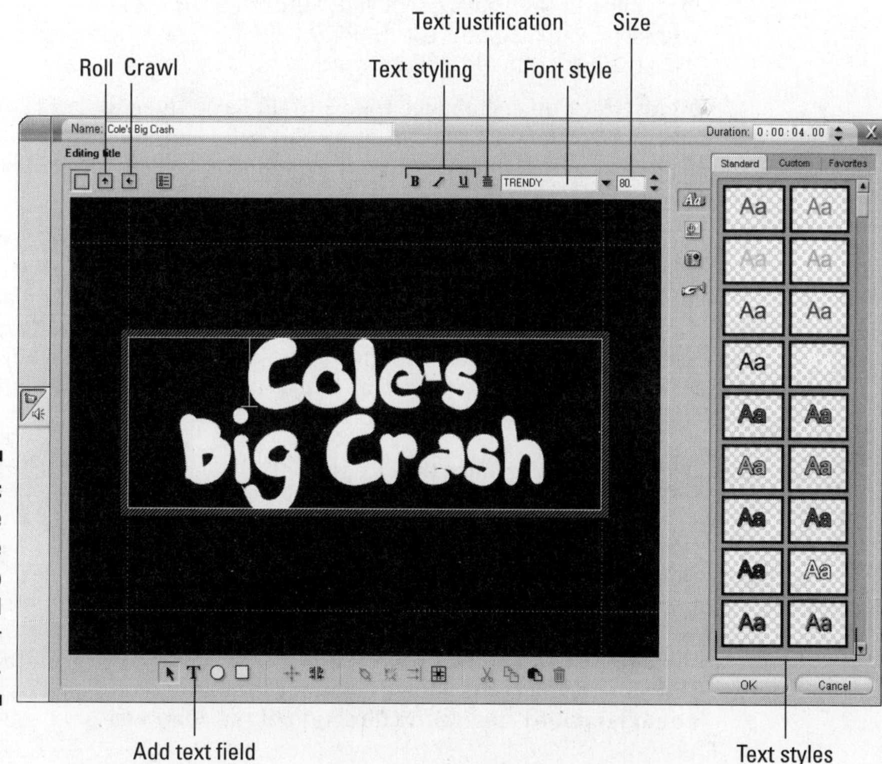

Figure 9-12: Use the Studio Title Editor to create and edit your titles.

When you edit titles in Studio, make sure that the words stay inside the title safe boundaries, which are the red dotted lines around the edges of the title editor window. Text that falls outside of those boundaries may be cut off when the movie is viewed on a TV screen.

To change the color of your text, click the Custom tab on the right side of the title editor screen. As you can see in Figure 9-13, this tab offers a variety of useful controls for text. The controls are divided into three sections: Face, Edge, and Shadow. Each section has a radio button for color, gradient, or transparent. If you want the element to be a solid color, choose the color radio button, and then click the color box next to it to open a color picker and choose a specific color. If you want the element to be a gradient (a color that gradually changes from one color at one side of the letter to a different color at the other side), click the gradient radio button and then click the gradient box next to it to specify the colors used in the gradient. If you want the element to be transparent, choose the transparent radio button.

The three sections of the Custom tab control separate elements of the title text:

- **Face:** This section controls the basic face of the text. You can choose a solid color, make the color a gradient, or make the text face transparent. Only choose the transparent radio button if the text has a thick, visible edge. You can also blur the appearance of the face using the slider control at the bottom of the Face section.
- **Edge:** Here you can change the color and size the edges of characters. Like the text face, edges can be a solid color, a gradient, or transparent. Choose the transparent option if you simply don't want the text to have an edge at all.

  Adjust the upper slider in the Edge section to change the size of the edge. Move the slider right to make the edge thicker, or move it left to make the edge thinner. Remember, very thin lines will flicker when the movie is viewed on a TV screen, so try to make the edge at least three pixels thick. An indicator to the right of the size slider tells you the current thickness in pixels. To blur the appearance of the text's edge, adjust the lower slider in the Edge section.
- **Shadow:** Add — and color — a shadow for your text, and control the direction of the shadow using the round direction control. Shadows can help make text easier to read over some backgrounds. Like other text elements, shadows can be a solid color, a gradient, or transparent. The upper slider in the Shadow section changes the apparent distance between the text and the shadow, and the lower slider blurs the shadow slightly. Generally speaking, the farther a shadow is beneath the text, the more you should blur the shadow. The radial control underneath the sliders lets you control the direction of the shadow. Click a radio button around the A to change the direction of the shadow.

When you're done editing your title, click OK in the lower-right corner of the Title Editor to close it and return to the Studio timeline.

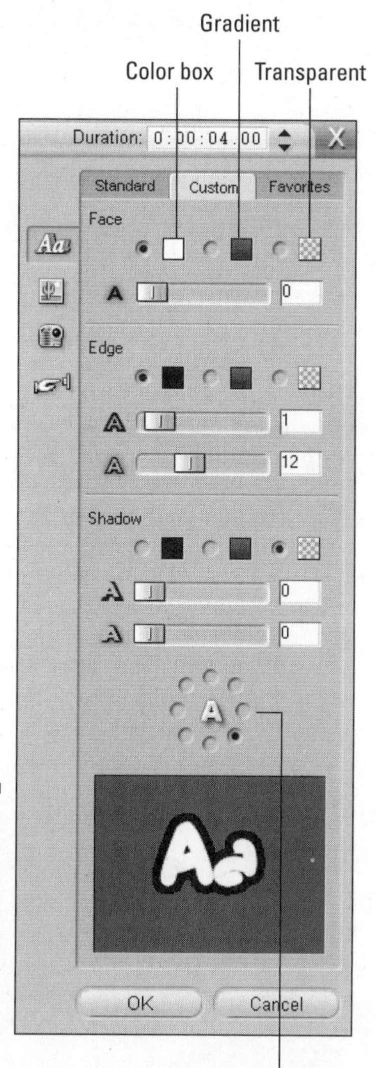

Figure 9-13: Change the color and appearance of text with these controls.

As you play with the Title Editor a bit, you'll probably find that creating titles from scratch takes a while. That's why I normally use one of the ready-made text styles (on the Standard tab of the Title Editor) as a starting point.

Changing backgrounds

If you've read this chapter from the beginning, you've already heard me preach about how white words over a black background are easier to read than any other style of title. But face it: Plain black backgrounds are a little, er, boring. If you want to spice up the appearance of your full-screen titles a bit, use the Title Editor to add a background. Here's how:

1. **Double-click a full-screen title in your timeline to open it.**
2. **Click the Backgrounds button to show the background controls.**
3. **To change the color of a solid background, click the background color radio button, and then click the color box to choose a new color.**

   You can also choose the gradient or transparent radio buttons if you want.
4. **To use a background image instead of a solid color or gradient, click the Background Picture radio button, and then click a picture in the list.**

In Figure 9-14, I have chosen a winter-themed background for my full-screen title. If you have a picture you'd like to place in the background, click the Pictures button, and then click the Folder icon in the pictures list to browse to the picture file on your hard disk. You can then click-and-drag the picture onto the title and position it anywhere you'd like. (For more on using still pictures in your movies, see Chapter 12.)

Figure 9-14: Studio comes with some nice background images for your titles.

Creating Titles in iMovie

Apple prides itself — and rightfully so — on providing computers and software that are easy to use. Adding titles to your iMovie project could hardly be easier. To create a title, follow these steps:

1. **If you're creating an overlay title, click the video clip in the timeline that will appear behind the title.**

   This operation ensures that the proper clip appears in the preview window as you edit and preview your title. If you're currently working in storyboard mode, I recommend that you switch over to the timeline by clicking the Timeline button, which appears to the lower-left of the monitor. If you're creating a full-screen title (one that is over a black background instead of superimposed over a clip), don't worry about selecting a clip in the timeline: Just go straight to Step 2.

2. **Click the Titles button above the timeline.**

   The iMovie title designer appears.

3. **Type the words for the title in the text boxes near the bottom of the title designer (as in Figure 9-15).**

4. **Choose a title style for your title from the Titles list (not to be confused with the brand of golf ball).**

   Most of the listed titles are for moving titles. When you click one, a preview of the motion appears in the preview window at the top of the title designer. If you want a static title that doesn't move, choose a Centered, Stripe, or Subtitle title.

5. **If the movie will be output to videotape or DVD for viewing on conventional TVs, make sure the QT Margins option is *not* checked.**

   The QT option allows titles in movies destined for the Web to use a little more screen space, but this isn't recommended for movies that will be shown on TVs. The problem is *overscan*, which I described earlier in the section, "Making Effective Titles."

6. **If you want to create a standalone title with a black background, place a check mark next to the Over Black option.**

7. **Click the Color box to choose a new color for the title.**

8. **Use the Font menu to choose a new font face, and adjust the size of the title using the size slider.**

9. **When you're ready to add the title to the project, drag the title from the Titles list and drop it into the timeline, just before the clip over which you want the title to appear.**

   The title will appear in the timeline. If you created an overlay title, it will be automatically superimposed on the video clip before which you placed the title. If you choose the Over Black option in Step 6, the title will be a standalone clip in the timeline with a black background.

If you decide to edit a title you created earlier, select the title in the timeline, make your changes, and then click the Update button at the top of the title designer window.

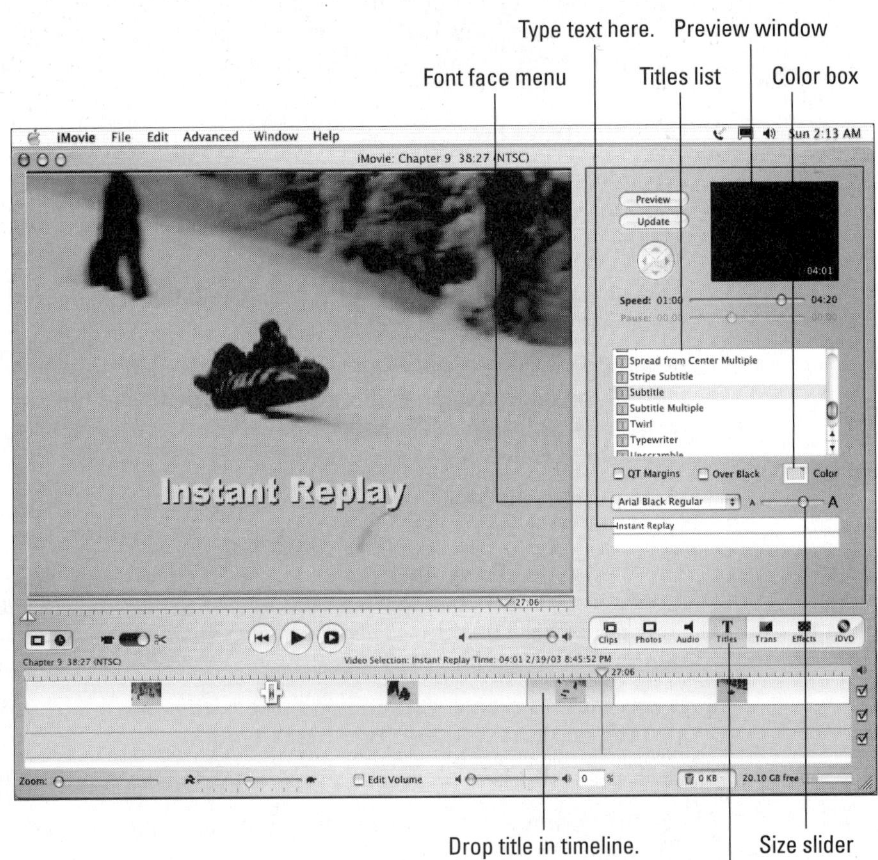

Figure 9-15: iMovie's title designer is easy to use.

Changing the Length of Titles

After you've added some titles to your movie project and previewed them a couple of times, you'll probably find that you need to change how long some of them appear. Does the title flash by so quickly that you don't have time to read it? Does the title seem to linger a few seconds too long? Changing the length for titles is pretty easy, but you must be viewing the timeline in your editing program, and not the storyboard. As I describe in Chapter 8, the timeline is where you perform "fine-tuning" edits such as changing the length of a title.

If you want to change the length of a title, you can do it in a couple of ways:

- In almost any video-editing program, hold the mouse pointer over an edge of the title in the timeline, and then click-and-drag the edge back and forth to make the title longer or shorter. This method is quick and easy but not very precise.
- In Pinnacle Studio, double-click the title to open the Title Editor. Change the time listed in the Duration field near the top of the Editor. You can either type a new number in the Duration field, or use the arrows next to the Duration field to increase or decrease the duration.
- In iMovie, move the Speed slider left or right in the title designer window.

In addition to altering the duration of your title, you can change which clip your overlay titles appear over. Just drag the titles left or right in the timeline if you want to change when they appear.

Chapter 10

Working with Audio

In This Chapter

- Understanding audio tracks
- Adjusting volume levels with rubberbands
- Adding sound effects to your movie
- Working with SmartSound

As we rush to trim video clips, create transitions, and add some cool special effects to our movies, it's easy to overlook the audio portion of the project. But the audio portion of a movie is nearly as important — or *as* important, depending on whom you ask — as the video picture.

In Chapter 7, I show you how to record better audio and import it into your computer. This chapter shows you how to make good use of that audio in your movie projects. Most movie-editing programs also come with built-in sound effects and other audio tools, so I show you how to use those as well. Finally, I also help you choose effective musical soundtracks for your movies.

If you don't have your own material to work with, you can still follow along with the steps in this chapter using the Chapter 10 sample clips found on the CD-ROM that accompanies this book. (See Appendix A for more on accessing and using the companion CD-ROM.)

Using Audio in a Project

Most movie-editing programs follow similar patterns. They all use storyboards and timelines for assembling the project, and most programs have similar windows for organizing and previewing clips. You'll also find that most programs have a lot in common when it comes to editing audio. For example, even the most affordable editing programs usually have separate audio tracks in the timeline for background music, narration (or sound effects), and the audio that accompanies the video clips in the timeline. (Audio tracks are described in greater detail in the next section.)

Many movie-editing programs can also show audio waveforms. A *waveform* is a line that graphically represents the rising and falling level of sound in an audio clip. Figure 10-1 shows the waveform for some narration I recorded in Pinnacle Studio. Waveforms are useful because they allow you to edit your audio visually, often with pinpoint accuracy. By looking at the waveform, you can tell when loud sounds or extended periods of quiet occur.

Although many movie-editing programs use audio waveforms, Apple iMovie isn't one of them. You can still edit audio (as I show later in this chapter); you just won't be able to do it visually in iMovie. Fortunately there are plenty of other programs that can display audio waveforms — including Apple Final Cut (both the Express and Pro versions), Adobe Premiere, Pinnacle Studio, and Windows Movie Maker.

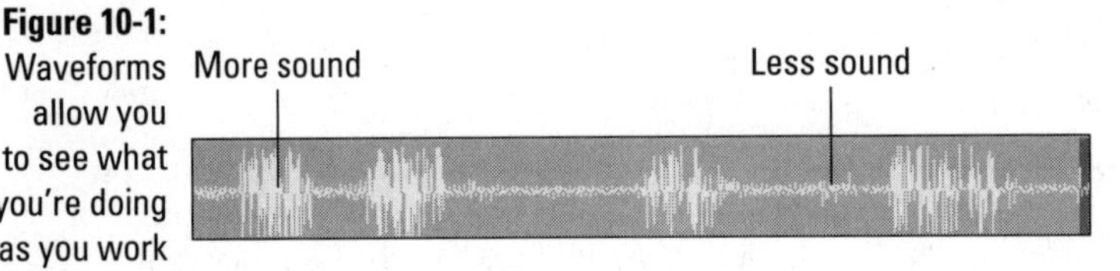

Figure 10-1: Waveforms allow you to see what you're doing as you work with audio.

Understanding audio tracks

A movie program can play several sources of audio at once. For example, while you hear the audio that was recorded with a video clip, you may also hear a musical soundtrack and some narration that was recorded later. When you're working on a movie project in your editing software, each of these unique bits of audio would go on its own separate audio track in the timeline.

Most editing programs provide audio tracks for main audio (the audio that was recorded with a video clip), music, and narration. More advanced editing programs offer you many more audio tracks that you can use any way you see fit. Adobe Premiere, for example, can provide up to 99 separate audio tracks in the timeline. Although it's difficult to imagine anyone actually *needing* that many audio tracks, having too many is better than not having enough.

Pinnacle Studio provides three separate audio tracks, as shown in Figure 10-2. To lock a track in Studio, click the track header on the left side of the timeline.

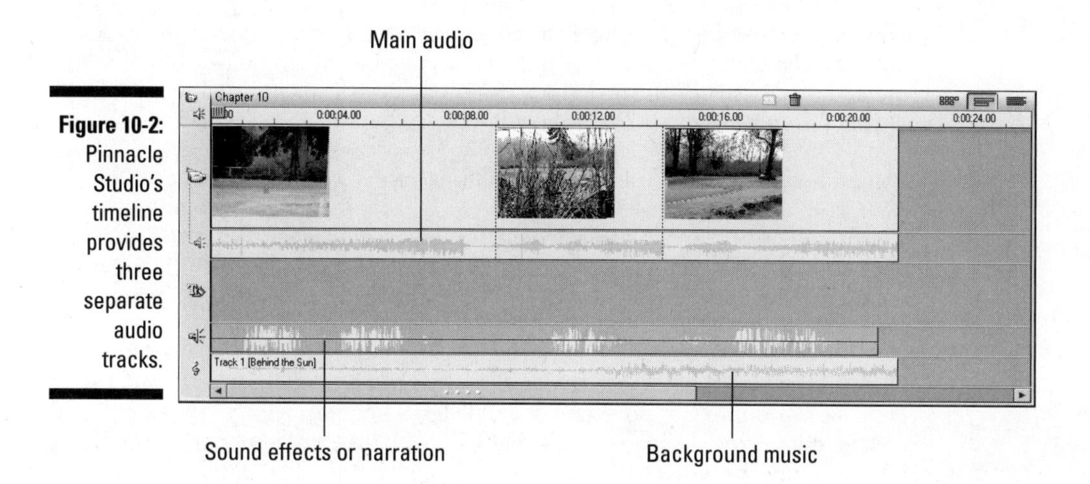

Figure 10-2: Pinnacle Studio's timeline provides three separate audio tracks.

Apple iMovie handles audio tracks a little differently, but you still have essentially three audio tracks to work with. The main audio track is actually hidden inside the video track. Two other audio tracks handle sound effects and background music. You can extract audio from video clips if you want (simply select the clip in the timeline and choose Advanced⇨Extract Audio), but doing so causes the main audio to take up one of the other two audio tracks. Figure 10-3 shows iMovie's audio tracks. You can enable or disable audio tracks by using the check boxes on the right side of the timeline. This is helpful during editing when you want to hear just one or two audio tracks at a time.

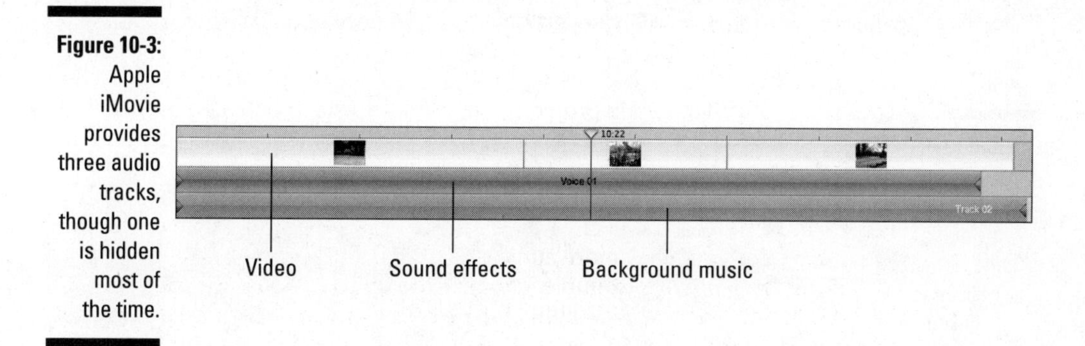

Figure 10-3: Apple iMovie provides three audio tracks, though one is hidden most of the time.

Adding audio to the timeline

Adding audio to the timeline is pretty easy. Record and import your audio as described in Chapter 7, and then simply drag-and-drop it to an audio track.

You can move clips by dragging them left or right in their respective tracks, and you can trim them by dragging on the edges. In fact, you'll find that editing audio tracks is a lot like editing video tracks — you can use the same basic editing techniques that I describe in Chapter 8.

Sometimes you'll want to edit main audio independently of the video clip with which it is associated. In iMovie, first extract main audio from the video clip by selecting the clip and choosing Advanced⇨Extract Audio. In Studio, click the track header on the left side of the timeline to lock either the main video or main audio track. When the main video track is locked, you can edit the main audio track without affecting clips in the video track.

One nice feature of iMovie is that it allows you to link audio clips to video clips. For example, suppose you've recorded some narration to go along with a video clip. If you decide to move that video clip to a different part of the timeline, you have to move the narration clip as well. If the two clips are linked, moving one will also move the other. To link an audio clip to a video clip in iMovie, follow these steps:

1. **Position the audio and video clips that you want to link in the timeline.**
2. **Move the play head to the beginning of the audio clip.**
3. **Click once on the audio clip to select it.**
4. **Choose Advanced⇨Lock Audio Clip at Playhead.**

A yellow pin appears on both the audio and video clips, as shown in Figure 10-4. If you move the video clip to a different place in the timeline, the linked audio clip moves with it. To unlink the clips, select the audio clip and choose Advanced⇨Unlock Audio Clip.

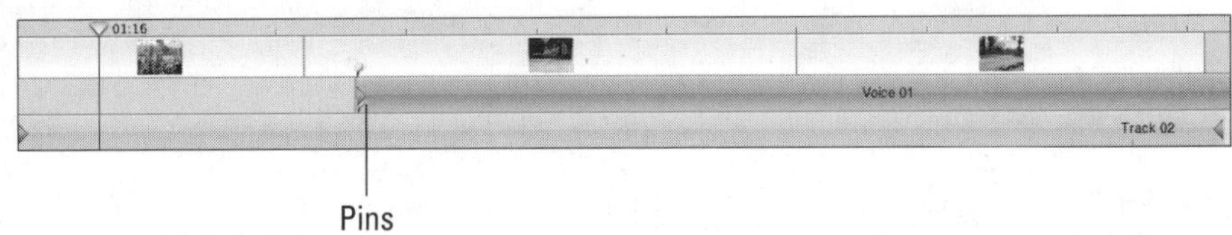

Figure 10-4: Yellow pins indicate that the audio and video clips are linked.

Adjusting volume

Perhaps the most common thing anyone does to audio tracks is adjust the volume. As you preview your project, you may notice that the background music seems a little too loud or the narration isn't loud enough. You may also have sounds in the main audio track that you want to get rid of altogether, without affecting the rest of the audio clip. Virtually all video-editing programs allow you to adjust volume in two different ways:

- You can adjust the overall volume of an entire clip or track.
- You can adjust volume dynamically within a clip, making some parts of the same clip louder and some parts quieter.

If you don't have audio clips of your own to work with, load the Chapter 10 sample clips from the CD-ROM that accompanies this book. Place all three scenes in the timeline in sequential order.

To begin adjusting volume in iMovie, place a check mark in the Edit Volume check box at the bottom of the timeline. In Studio, click the clip you want to modify and then choose Toolbox➪Change Volume so that Studio's audio toolbox appears above the timeline.

Each program displays audio rubberbands across audio clips. *Rubberbands* (see Figure 10-5) aren't just for holding together rolled up newspapers or your hairdo; in video programs, they show you the volume for an audio clip, and they allow you to make dynamic adjustments to the volume throughout the clip.

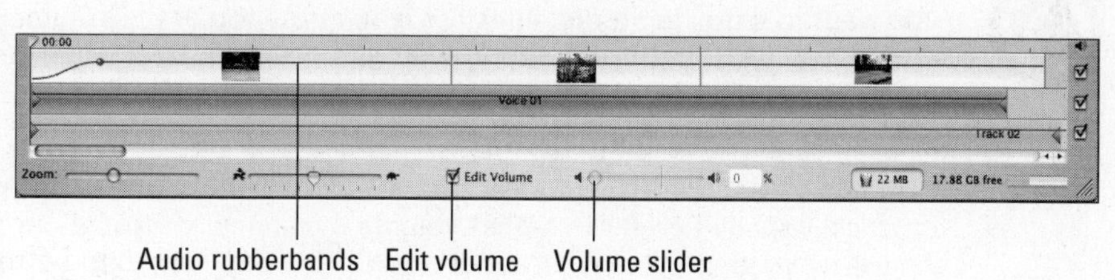

Figure 10-5: Rubber bands make it easy to adjust volume in your audio clips.

Adjusting overall volume

Modifying the overall volume for an audio clip or a whole track is pretty simple, but the procedure varies a bit depending on which program you are using.

In Apple iMovie, follow these steps:

1. **Click a clip to select it. (To adjust multiple clips simultaneously, hold down the ⌘ key as you click each clip.)**

   Be careful not to click the purple rubberband line. If you accidentally click the rubberband and a dot appears on the line, press ⌘+Z to undo the change.

2. **With the desired clip(s) selected, adjust the Volume slider leftward to reduce volume or right to increase volume.**

   As you adjust the slider (shown in Figure 10-5), you see the rubberband lines move up or down.

Pinnacle Studio provides a variety of volume controls in the audio toolbox, as you can see in Figure 10-6. To open this toolbox, click an audio clip in the timeline and choose Toolbox➪Change Volume. The toolbox contains a separate set of controls for each of the three audio tracks:

- **Left:** Main audio track
- **Middle:** Sound effects/narration track
- **Right:** Background music track

To adjust the overall volume for an entire track, turn the knob at the top of the track's volume controls. As you turn the knob, you see the blue audio rubberband line for the entire track move up or down in the timeline. If you want to adjust the volume for a specific clip, place the play head at the very beginning of that clip in the timeline and move the volume slider for that track up or down. Then place the play head at the end of the clip, and adjust the volume slider back to the middle to restore the volume of subsequent clips in the timeline.

You can also mute whole tracks in the timeline if you want. In iMovie, simply remove the check mark from the right side of the timeline next to the track you want to mute. In Studio, click the icon at the top of the volume controls (see Figure 10-7) in the audio toolbox to mute a given track.

Adjusting volume dynamically

Believe it or not, I very seldom adjust the overall volume of an entire clip or track. Usually I prefer to adjust volume dynamically throughout the clip. Adjusting volume dynamically allows me to fine-tune audio to better match other things that are going on in the project. I adjust volume dynamically at times like these:

- **Narration is about to begin:** I may reduce the volume of background music a bit so that the spoken words are more easily heard.
- **The sound changes between video clips:** Such a change often sounds abrupt or harsh. I can reduce that by fading audio in at the beginning of the clip, and fading it out at the end. Studio has handy Fade In and Fade Out buttons (see Figure 10-6), which automatically make fade adjustments to the audio rubberbands for a clip.

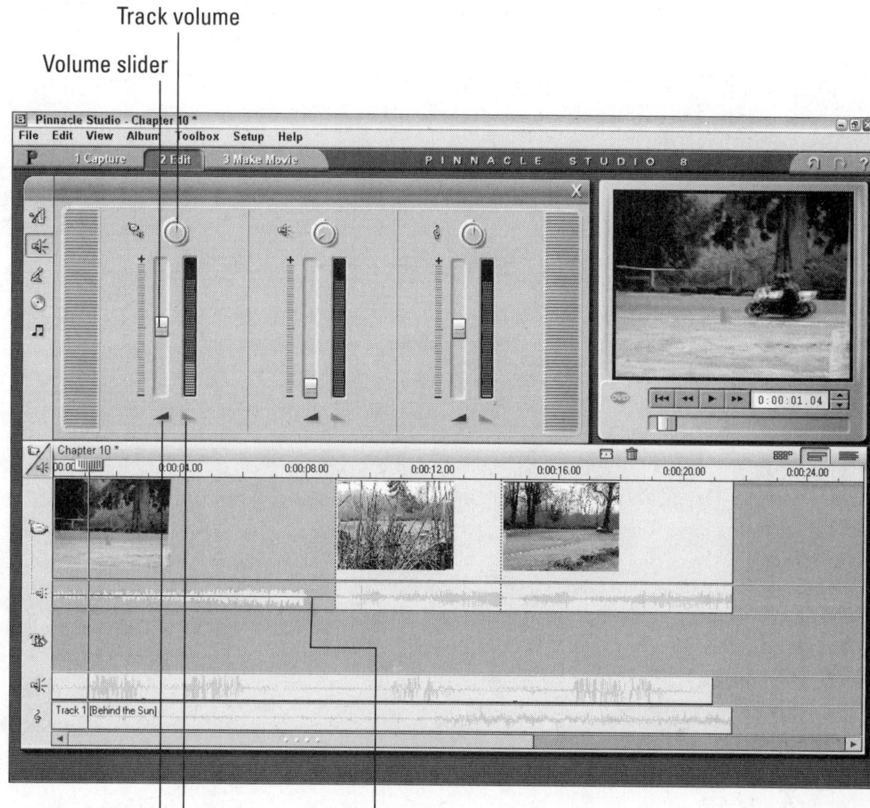

Figure 10-6: Use Studio's audio toolbox to adjust volume.

- **I can eliminate unwanted sounds from a clip:** For example, the Scene 1 sample clip from the companion CD-ROM shows a motorcycle passing by the camera on a racetrack. But the beginning of the clip has the sound of another motorcycle that is actually behind the camera. I can eliminate the sound of that other off-camera motorcycle by dynamically adjusting volume as I've done in Figure 10-7.

Dynamic volume adjustment is where audio rubberbands really come in handy. Click once on a rubberband for one of your audio clips. When you click the rubberband, you'll notice that a little dot appears. You can click-and-drag that dot up or down to adjust the volume of the clip. Not only can you place as many dots as you can squeeze onto a rubberband, you can also move those dots around on the rubberband to make constant adjustments throughout the clip. Like real rubberbands, audio rubberbands are stretchy and can be moved quite a bit, but unlike real rubberbands, they don't snap back when you stop moving them. Play around a bit with the rubberbands to see just how fun volume adjustments can be!

Figure 10-7: Use rubberbands to adjust volume dynamically throughout an audio clip.

Those little dots on audio rubberbands act like pins to hold the bands in place. If you want to get rid of a dot, click-and-drag it off the top or bottom of the clip. Twang! The dot will disappear and the rubberband will snap back into place.

Working with sound effects

Sound effects can really separate a good movie from a great movie. In Chapter 7, I suggested that if a picture is worth a thousand words, sometimes a sound is worth a thousand pictures. Consider how various sound effects can affect the mood of a scene:

- A quiet room somehow seems even quieter if you can hear the subtle ticking of a clock.
- Applause, cheering, or laughter suggest how the viewer should feel about an event in the movie.

- Chirping birds suggest peace and serenity.
- Footsteps make movement seem more real, even when the feet are not shown in the video image.

Those are just a few examples, but you get the idea. You can create and record your own sound effects if you want, and in Chapter 17, I offer some tips for doing exactly that. Fortunately, most video-editing programs now come with ready-to-use libraries of common sound effects to enhance your movie projects. The next two sections show you how to find and use the sound effects in iMovie and Studio.

Using sound effects in Apple iMovie

To view a list of sound effects in iMovie, click the Audio button above the timeline and choose iMovie Sound Effects from the menu at the top of the audio pane. A list of sound effects appears (as shown in Figure 10-8), including standard sound effects as well as special Skywalker Sound Effects that were new with iMovie 3. Skywalker Sound, as you may know, is the brainchild of George Lucas of *Star Wars* fame, and the sound effects included with iMovie are very high-quality.

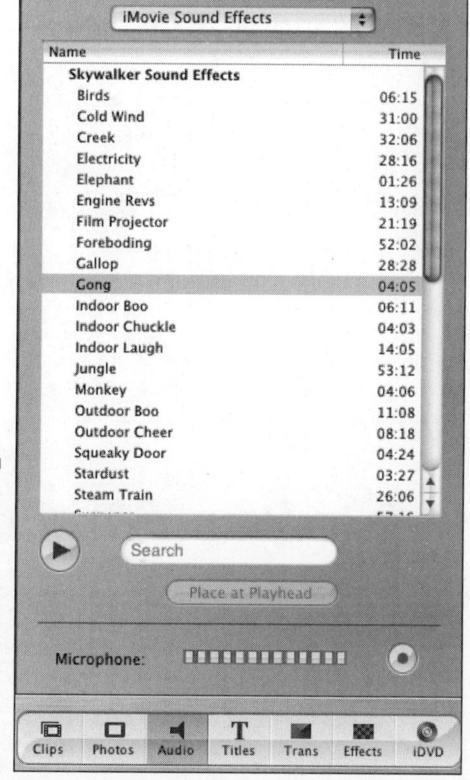

Figure 10-8: Apple iMovie includes a collection of high quality Skywalker Sound Effects.

To preview a sound effect, choose it from the list and click Play in the audio pane. To use a sound effect in your project, simply click-and-drag it to an audio track in the timeline. You can trim and adjust the volume of sound effects just as you would any other audio clip.

Using sound effects in Pinnacle Studio

Pinnacle Studio comes with a diverse collection of sound effects. To view a list of them, click the Show Sound Effects tab on the left side of the album or choose Album ➪ Sound Effects. When you view the Sound Effects tab of the album, you may not actually see a list of effects, particularly if you've used this tab to import MP3 audio or other sounds not directly connected with Pinnacle Studio. If sound effects aren't shown, you'll have to click the folder icon at the top of the album and browse to the following folder on your hard drive:

```
C:\Program Files\Pinnacle\Studio 8\Sound Effects
```

In that folder, you will see a selection of 13 subfolders, each of which contains a category of sound effects. For example, if you want to find chirping birds, double-click the Animals folder and then click Open. A list of animal sound effects will appear in the album (as shown in Figure 10-9). To preview a sound effect, simply click it in the album. To use it in your movie project, click-and-drag it to the sound effects/narration track in the timeline.

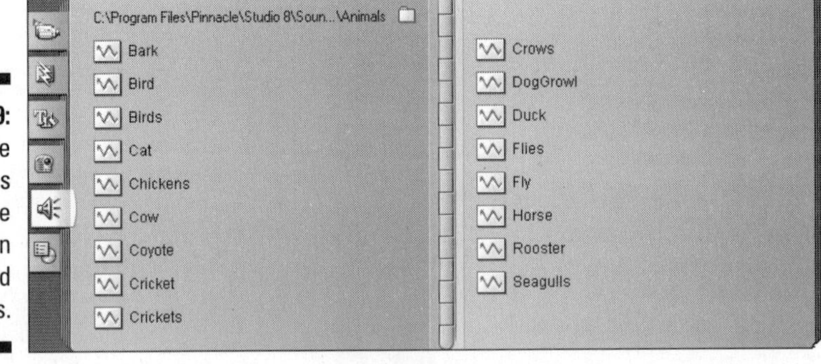

Figure 10-9: Pinnacle also comes with a wide collection of sound effects.

Adding a Soundtrack to Your Project

Background music for movies is nothing new. Even before the day of "talkies," silent movies were traditionally accompanied by musicians who played in the theater as the movie was shown. I'm sure you're eager to use music in some of your movie projects as well. The next two sections show you how to add soundtrack music to your movies.

Adding music from a CD

In Chapter 7, I show you how to import music from audio CDs or MP3 files. After you have inserted an audio file into your editing program, you can add the file to the music track in your timeline. The procedure varies depending on which editing program you are using:

- In Apple iMovie, use iTunes (described in Chapter 7) to import music from audio CDs into your iTunes library. Then click the Audio button in iMovie, choose iTunes from the menu at the top of the audio pane as shown in Figure 10-10, and click-and-drag a song from your iTunes library to an audio track on the iMovie timeline.
- In Pinnacle Studio, place a music CD in your CD-ROM drive and choose Toolbox➪Add CD Music. The Add CD Music toolbox appears. Choose the audio track that you want to use in the Track menu, and then click Add to Movie. Close the Add CD Music toolbox, and then click Play under the preview window to play your timeline. The first time you play the timeline, Studio will import the music from the CD. After the music is imported, you can remove the CD from your CD-ROM drive.

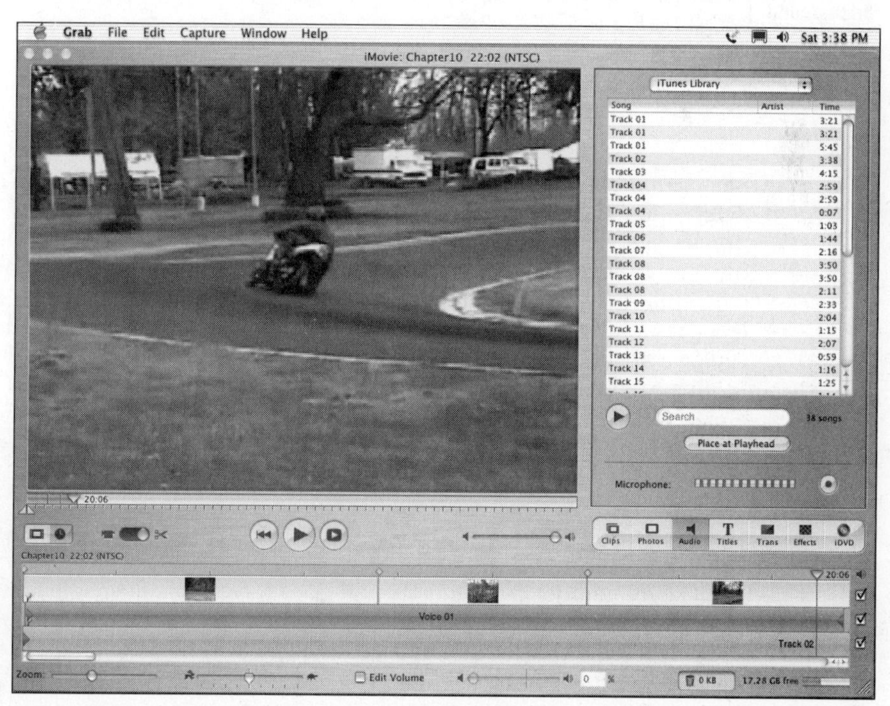

Figure 10-10: iMovie lets you import music directly from your iTunes library.

Generating background music with SmartSound

Pinnacle Studio comes with a tool called SmartSound, which can automatically *generate* music in a variety of styles to match your project. To generate some music from SmartSound, open a project in Studio and choose Toolbox⇨Generate Background Music. The audio toolbox appears, as shown in Figure 10-11. (If you did not install the SmartSound files on your hard disk during installation, you will need to place the Studio program disc in your CD-ROM drive.)

Figure 10-11: SmartSound enables you to create varying styles of background music for your movie.

SmartSound offers music in a variety of styles. Each style includes several songs; most have a few different styles available. Click the Preview button to preview a style, song, and version. When generating background music, figure out approximately how long you want the musical piece to play. For example, when the three Chapter 10 sample clips are placed in the timeline, their total length is a little more than 21 seconds. Adjust the duration of the music by typing a new time in the Duration box (in the upper-right corner of the audio toolbox). When you click Add to Movie, SmartSound automatically generates a piece of music in the chosen style, song, and version, and it plays for approximately the duration you chose. You can also name the selection if you want, using the Name field at the top of the toolbox.

One advantage of SmartSound's automatically generated music is that you can use it without paying royalties every time someone views your movie. Even so, remember that SmartSound does have *some* licensing restrictions. They're worth reviewing before you use them in a movie you plan to show to the public. For SmartSound license details, click the SmartSound button in the Studio audio toolbox.

Chapter 11

Advanced Video Editing

In This Chapter

- Using video effects in iMovie
- Applying video effects in Studio
- Stepping up to advanced video editing programs

If you're brand new to this whole movie-editing thing, having the capability to add transitions, titles, and musical soundtracks to your videos probably seems pretty cool. But transitions, titles, and sounds are just the beginning of what you can do. Even the most basic video-editing programs these days provide some special-effects features. Slightly more advanced programs — such as Adobe Premiere, Apple Final Cut Express, or Pinnacle Edition — offer truly astounding special-effects capabilities that even moviemaking professionals could barely afford to dream about just a few years ago.

This chapter shows you how to use some of the special-effects features already available to you with the software you're using right now. I also give you a taste of the editing power that may be available to you if you step up to a more advanced video-editing program.

Using Video Effects in iMovie

When I mention "special effects," what's the first thing that pops into your mind? You're probably envisioning spaceships soaring between the stars, giant monsters destroying a city, or a superhero soaring unassisted through the sky. Such effects are indeed special, but they only scratch the surface of how effects can be used in movies. Special effects can

- Transport the viewer to a different place and time.
- Show subjects that would be otherwise impossible or too expensive to photograph.
- Change the mood or feel of a piece of video.
- Add visual excitement or a sense of the exotic.

Special effects aren't all about spaceships and monsters. Consider the scene in Figure 11-1. I have used a Mirror effect in iMovie to create a kaleidoscopic image. This effect would look right at home in a music-video-style project.

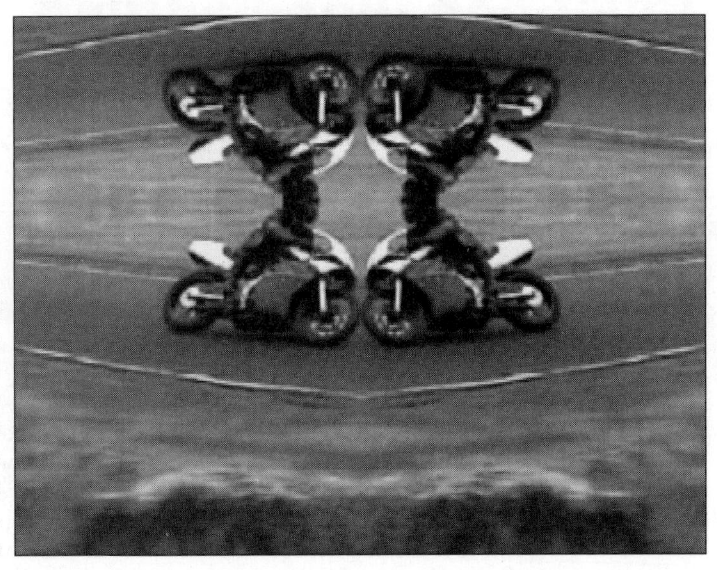

Figure 11-1: Special effects can make regular clips seem more exotic.

Reviewing iMovie's effects

For a free program, Apple's iMovie sure comes with a lot of great features, including some special effects that can quickly transform the look and feel of your video clips. Besides the effects that come with iMovie, you can add effects by installing plug-ins from Apple and third-party software vendors. (Appendix C shows you how to install plug-ins in iMovie.) To view a list of iMovie's effects, click the Effects button. The Effects pane appears, as shown in Figure 11-2. To preview an effect, just click it in the list of effects. A small preview of the effect appears at the top of the Effects pane, using whatever video is located at the current position of the play head in the timeline.

Some iMovie effects are featured elsewhere in this book. Chapter 1, for example, shows you how to use the Aged Film effect to make your video look like it was shot on film many years ago. Chapter 8 shows you how to modify light and color in your clips by using some more iMovie effects. Other effects include

- **Earthquake:** As the name implies, this effect shakes the video image as if an earthquake were happening. Use a short earthquake effect combined with the sound of a large thud when someone falls down in your video.

- **Electricity:** This one has a shocking effect on your video by creating a continuous lightning bolt in your video. Use this effect in conjunction with the Electricity sound effect in the Audio pane (see Chapter 10 for more on working with sound effects).
- **Fairy Dust:** This effect looks more like a burning fuse to me, but the folks at Apple call it "Fairy Dust," so who am I to argue?
- **Flash:** The Flash effect makes the video image flash periodically. Combine the Flash effect with the Earthquake effect to create a more disorienting scene, or combine it with the Aged Film effect to make your video look *really* old and deteriorated.
- **Fog:** Add some mystery to your scene with the Fog effect. This effect is a classic example of an image that is often easier to edit into your video via software than to actually photograph. Use the Fog and Black & White effects together for a *film noir* look.
- **Ghost Trails:** Slight ghost trails are added to moving objects in the video image. Here's another effect that can be used to make a scene look disorienting.
- **Lens Flare:** Lens flares are bright spots or streaks that occur when the sun or another bright light source reflects on the camera lens. Chapter 4 shows you how to *avoid* lens flares when shooting video, but sometimes you may want to add a lens flare to make a mostly-imaginary scene look "photographed." Professional moviemakers often add "fake" lens flares to animated or computer-generated images to make the scene look more realistic.
- **Letterbox:** Simulate that Hollywood movie "widescreen" look with this effect, which puts a black bar across the top and bottom of the video image. Just keep in mind that the top and bottom of your video image will be cut off.
- **Mirror:** iMovie actually offers two Mirror effects that turn half your video picture into a mirror image of the other half. One Mirror effect reflects only horizontally; the other can reflect both horizontally and vertically. iMovie's Mirror effect is demonstrated in Figure 11-1.
- **N-Square:** Ever wonder how a housefly would see your video? The N-Square effect may show just that. This effect divides the screen into squares and copies the video clip to each square. You can adjust the number of squares from four up to 64 as shown in Figure 11-3.
- **Rain:** Add some fake weather to your scene with this effect. The fake rain drops added to a video clip by this effect show up much better onscreen than real rain, yet the scene still looks realistic. Use this effect in conjunction with the Rain sound effect in the Audio pane.
- **Soft Focus:** Simulate a soft filter on the camera with this effect. The Soft Focus effect works well in dream sequences, or to cover up blemishes on a video subject.

Figure 11-2: iMovie provides various video effects in the Effects pane.

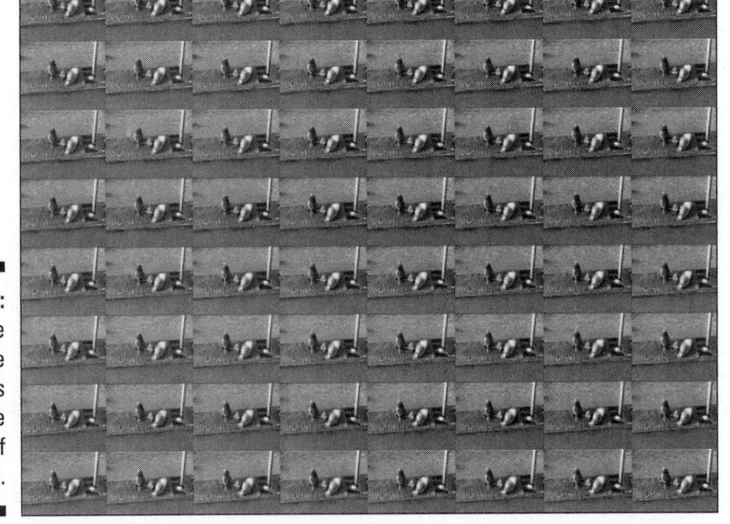

Figure 11-3: The N-Square effect gives a bug's-eye view of your video.

Customizing effects

Most effects can be customized to some extent. When you click an effect, controls relating to that effect should appear in the Effects pane, as shown in Figure 11-4. The exact controls that will be available vary depending on the effect, so some experimentation may be necessary. In Figure 11-4, I am adjusting the Virtix Flame effect, which is available as a free iMovie plug-in from Virtix (`www.virtix.com`). Slider controls at the bottom of the Effects pane let me control the height and appearance of the flames. Some other controls on the Effects pane are common to all effects. These include

- **Effect In:** This slider lets you control when the effect begins. If you adjust this slider to create a delay, the effect appears gradually as the clip plays. If you want the effect to apply to the whole clip, leave this slider set to 00:00.
- **Effect Out:** Like the Effect In control, this slider lets you end the effect before the clip ends.
- **Preview:** If the tiny little preview screen in the Effects pane isn't big enough to see the effect, click the Preview button. A short preview of the effect appears in the main iMovie viewer window, as shown in Figure 11-4.
- **Apply:** Click this to apply the effect to the currently selected clip in the timeline.

Figure 11-4: The Virtix Flame effect makes it look like these sea lions are in hot water.

Controlling effects with keyframes

When you apply an effect to a clip, you may want the effect to appear only for part of the clip, or you may want the effect to change as the clip plays. In iMovie, you can make an effect appear or disappear gradually. More advanced video-editing programs like Adobe Premiere and Apple Final Cut Pro control effects by using *keyframes*. A keyframe is simply a frame of the movie that you set as a landmark (as you would the points you create and adjust on audio rubberbands when you edit audio, as described in Chapter 10). In advanced video-editing programs, you can set as many keyframes as you like, and you can change the properties of effects at each keyframe.

For example, suppose you use a rain effect on a video clip, and you want the rain to increase and decrease in intensity throughout the clip. Several seconds into the clip, you can set the rain to be very heavy. A few seconds later, create another keyframe where the rain is very light. The software automatically adjusts the Rain between those keyframes to make the change appear gradual.

Using keyframes effectively take a while to learn, and if you use a program like Adobe Premiere, I recommend you get a book that more fully describes the features of that program. (Naturally I recommend *Adobe Premiere For Dummies,* published by Wiley Publishing, and written by yours truly.) But the keyframe feature does give you almost infinite control over the effects in your movies, so it's definitely worth seeking out if you plan to move up to a more advanced editing program.

After you apply an effect to a video clip, iMovie must render the effect. A thin progress bar will appear on the clip in the timeline. *Rendering* is a process where iMovie applies your effect to the clip and creates a new video file on your hard disk that incorporates the change. If you decide later that you want to remove an effect from a clip, select that clip in the timeline, and then use the effect controls in the Effects pane to reduce the amount of the effect in the scene. You can't remove the effect completely, but most effects can be disabled using the slider controls at the bottom of the Effects pane.

Using Video Effects in Studio

Pinnacle Studio doesn't have built-in special effects, per se, but it does offer tools to let you modify the appearance of your video clips. I describe these tools throughout *Digital Video For Dummies*, but here's a summary of the tools and effects available in Studio:

- **Adjust color and lighting:** You can modify colors and lighting in video clips using controls in the Adjust Colors toolbox. To open this toolbox, select a clip in the timeline and choose Toolbox⇨Adjust Color. The toolbox appears above the timeline, as shown in Figure 11-5. Most of the time you won't want to make any extreme adjustments, but you can correct color problems in video images with a little fine-tuning.

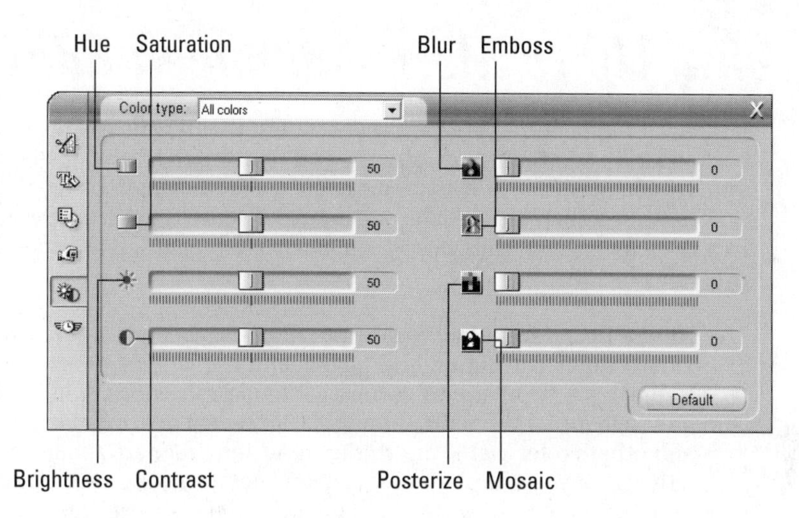

Figure 11-5: Studio's Adjust Colors toolbox lets you make basic adjustments to your video image.

If you're going for a black-and-white look in your movie, you don't need any special effects. Simply adjust the Saturation slider all the way to zero to remove color from the video image.

- **Blur:** As you might guess, the Blur slider blurs the video image. If you set the Blur slider to number one, it does a fair imitation of a soft filter. Any higher setting, though, and the image just looks, well, blurry.
- **Emboss, Mosaic, and Posterize:** These controls are of limited use, in my opinion, although they can be useful when you're creating a custom background image (for a DVD menu, for example):
  - **Emboss** gives the appearance that your video image has been embossed into plastic or metal.
  - **Mosiac** changes the image into a pattern of colored square blocks. As you adjust the slider farther to the right, the blocks become larger, further distorting the image.
  - **Posterize** reduces the number of colors in the image to create a cartoon-like appearance.
- **Vary playback speed:** Studio has a really nice toolbox for adjusting the playback speed of video clips. To open this toolbox, choose Toolbox⇨Vary Playback Speed. (I describe this feature in greater detail in Chapter 8.)
- **Overlay graphics:** You can use the title editor to lay simple graphics over a video image. Check out Chapter 9 to see an example that adds a flying saucer to an image.
- **Hollywood FX Pro:** Pinnacle offers Hollywood FX Pro, a tool that can create some advanced video effects. (See Chapter 12 for a detailed look at one way Hollywood FX can be used.) Hollywood FX Pro is available as a plug-in for Studio for $99.

Stepping Up to More Advanced Editing

If it hasn't happened to you yet, rest assured that it will: You want to do a certain kind of edit, but the software you're using right now just won't let you. Or you may find that some tasks are difficult or tedious to perform. Before you get mad and break a keyboard or something, consider upgrading to a more advanced editing program. Some well-regarded advanced editing programs include

- **Adobe Premiere (Mac/Windows):** Adobe kicked the affordable video editing revolution into high gear a few years ago with Premiere. It remains a popular choice among video professionals and hobbyists, although it has slipped in popularity in recent years as Final Cut Pro and others offer viable alternatives. Adobe Premiere retails for $549, although if you shop around, you may get a better deal by buying Premiere packaged with a high-quality video-capture card.
- **Avid Xpress (Mac/Windows):** Avid has been making broadcast video-editing software and equipment for years, and many professionals prefer the Avid Xpress video-editing software. And after years of spending tens (if not hundreds) of thousands of dollars for Avid editing workstations in the past, pros don't mind the $1699 retail price of Avid Xpress. For the typical amateur moviemaker, however, some of the more affordable solutions make more sense.
- **Apple Final Cut Pro (Mac):** Quickly becoming one of the most popular editing programs among pros and dedicated hobbyists, this program, which retails for $999, can do it all.
- **Apple Final Cut Express (Mac):** If Final Cut Pro's price tag is out of your reach, consider the $299 Express version. Unless you're editing video professionally for broadcast TV or Hollywood movie studios, this is a better, much more affordable choice.
- **Pinnacle Edition (Windows):** Pinnacle definitely knows video. Whether you're a broadcast professional or DV newbie, Pinnacle has software and hardware worth considering. Edition is Pinnacle's answer to Adobe Premiere and other high-end video editors. Pinnacle Edition DV retails for $699 with a FireWire card included, although if you already own Pinnacle Studio, you can upgrade to Edition for a lot less (at this writing, Pinnacle advertised an upgrade special of just $199).
- **Sonic Foundry Vegas (Windows):** Vegas from Sonic Foundry has been quietly building a strong reputation among video editors, many of whom prefer Vegas to Adobe Premiere for Windows-based editing. The Vegas software retails for $699, though Sonic Foundry occasionally has special offers.

Chapter 19 provides a feature-by-feature comparison of these and other editing programs. As you can see, if you want a professional-style video editor, you're going to pay a few hundred dollars, if not more. So what do you get for your money?

Although you may expect advanced editing programs to be more complex, I actually find them easier to use. For example, consider the process of capturing video from your camcorder. In Apple iMovie and Pinnacle Studio, you have to be quick with your mouse and buttons to start and stop capture. If you want to capture a lot of different scenes from a tape, this process can grow tedious because you have to sit there in front of your computer manually starting and stopping capture for each piece of video. And because you probably won't actually capture a piece of video until you've reviewed it at least once, you'll spend a lot of time rewinding the same video over and over.

Most high-end editors simplify the capture process with a feature called *log and capture*. The log and capture feature allows you to create a log of clips that you want to capture. As you review the videotape and identify portions that you want to capture — a task you would undertake with any video-editing program — you simply log the timecode at the in point (beginning) and out point (end) of each piece of video you want to capture. In Adobe Premiere (refer to Figure 11-6), you do this by clicking Set In at the beginning of the section, and then clicking Set Out at the end of the section. You can even type the timecode in the Set In and Set Out fields manually if you wish. When you've specified an in point and an out point, click the Log In/Out button. This adds the current in and out points to a log. The Batch Capture window in Figure 11-6 serves as Adobe Premiere's log. You can log as many sections from a tape as you want in a single log. In Figure 11-6, I have logged five separate sections of video.

After you have logged all of the portions of video that you want to capture from a given tape, capturing the video takes just a single mouse click. In Premiere, click the Record button at the bottom of the Batch Capture window. The software automatically rewinds and cues the tape in the camcorder as needed and captures all the sections of video that you specified. The capture process may take a while, but because it all happens automatically, you can take a lunch break while it occurs. If you capture a lot of video the old fashioned way in Apple iMovie or Pinnacle Studio, you'll start to see why log and capture can be a big timesaver. Log and capture is just one example of the effort-saving features that advanced video-editing programs can offer.

All the editing programs listed here look and work very similarly because they were designed to imitate the professional video-editing workstations used for years by broadcast pros. One of the things I really like about programs like Premiere or Final Cut Pro is the flexibility of their timelines. You normally have an almost infinite number of tracks to work with — very helpful when you want to overlay video images on top of each other. Consider the Premiere timeline in Figure 11-7. It may look like a very complex project, but I was able to assemble it quickly. The timeline in Figure 11-7 currently has three video tracks and three audio tracks, but I can add more as I need them.

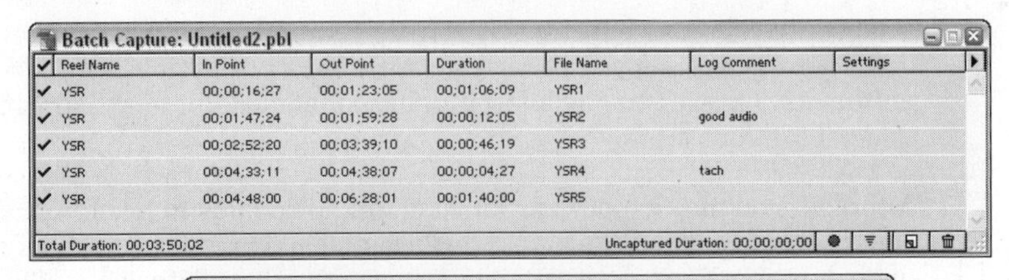

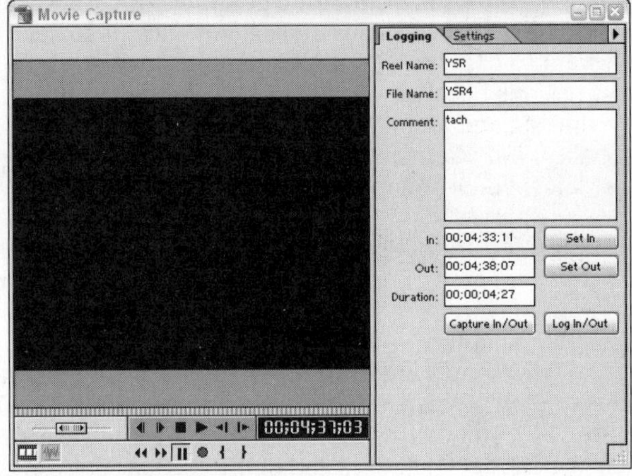

Figure 11-6: Adobe Premiere gives you more control over the video capture process.

Figure 11-7: Programs like Premiere may look complex, but they actually make video editing easier.

Special effects are also a lot easier to work with in advanced editing programs. For example, suppose I want to make a video clip appear in its own little box (like a picture-in-picture display found on some TVs) and move the box across the screen. In Premiere, opening a Motion Settings dialog box for the clip is a simple matter, as shown in Figure 11-8. Using simple point-and-click techniques with the mouse, I can draw a motion pattern for the clip and adjust zoom settings to make the clip bigger or smaller.

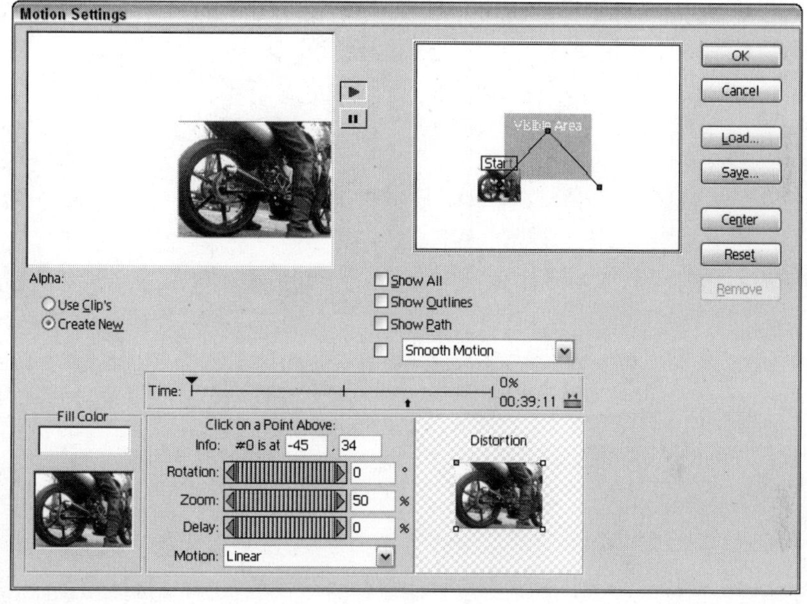

Figure 11-8: Animation and other effects are easy in advanced video-editing programs.

Advanced video editors can also be powerful business productivity tools. Corporate communications directors are finding that modern computers and software make in-house video production more feasible (in the past it was usually farmed out to a contractor). Businesses can use video editors to make

- Training and orientation videos
- Kiosk videos
- Marketing materials
- Presentation videos

One advantage of using video software for presentations is that once a presentation is exported to tape or DVD, you can show it almost anywhere. All you need is a TV and a VCR or DVD player. Although you lose some of the interactivity of, say, a PowerPoint presentation, you can still make a very professional-looking video using media of almost any type. Last year, for example, I used Adobe Premiere to develop a kiosk video to demonstrate some software. Using animation tools like the one shown in Figure 11-8, I could simulate the movement of a mouse pointer on the screen.

I can't possibly show you all the advantages that advanced video-editing programs can offer, but this chapter shows some possibilities. Programs like Final Cut Pro, Edition, and Premiere are so advanced that the movie magic you can create is limited only by your imagination.

Chapter 12

Working with Still Photos and Graphics

In This Chapter

- Using still images in movies
- Preparing a still image to be used in a movie
- Creating still pictures from video images

Webster's New World Dictionary tells us that the word *movie* is a contraction of the term "moving picture." When you view a moving picture, you are actually viewing a series of still pictures that flash by so quickly that they provide the illusion of motion.

Dictionaries, contractions, and illusions are all well and good, but take the "moving" out of moving pictures for a moment. Sometimes you'll want to include a still picture or graphic in a movie. Likewise, sometimes you'll want to turn an image from your movie into a still picture. Both of these things are pretty easy to do with modern video-editing programs, and this chapter shows you how.

Using Still Graphics in Movies

Although movies are all about motion, there are plenty of times you'll want to include still graphics in your video projects. The process of importing stills into your video-editing program is usually pretty simple, but before you start plopping stills into your timeline, you should take some important steps to make sure that your images will look right. If you don't take the time to prepare your stills, two main problems can result:

- Images may become distorted and look unnaturally stretched or squeezed.
- Colors might not display properly.

The following sections help you prepare and use still graphics in your movies.

Adjusting the image size

In Chapter 3, I mention the concept of aspect ratios in video. To briefly recap, the *aspect ratio* describes the basic shape of the video image. (Figure 3-3 in Chapter 3 shows the difference between two common aspect ratios.) Most video images have an aspect ratio of 4:3, which means the width and height of the image can be divided into a 4:3 ratio. Most of the sample clips on the CD-ROM that accompanies this book have an image size of 320 pixels wide by 240 pixels high. If you do the math, you'll see that 320x240 works out to a ratio of 4:3.

When you take a picture using a digital still camera, that picture usually has an aspect ratio of 4:3 as well. Image sizes that have a 4:3 aspect ratio include 640x480, 800x600, and 1024x768. Your computer's monitor also probably has a 4:3 aspect ratio unless you have a newer widescreen monitor. Most TVs still have a 4:3 aspect ratio as well. Because the 4:3 aspect ratio is so common, dropping an 800x600 digital photo into a DV-based video project should be easy, right?

Unfortunately, it's not always that easy.

Digital images are made up of *pixels*, which are tiny little blocks of color that, when arranged in a grid, make up a picture. Pixels can have different aspect ratios too. Digital graphics usually have square pixels, while video usually has rectangular pixels. The frame size of NTSC video is usually listed as 720x480 pixels. This does not work out to 4:3; it's actually 3:2. However, it still *appears* to have a 4:3 aspect ratio because NTSC video pixels are slightly taller than they are wide.

What does all this mean? Suppose you place an uncorrected digital picture into a video program, and then you export that program back to tape. When you view the movie on a TV, the still image will appear distorted. To correct this, you need to start with a digital photo that has an aspect ratio of 4:3, and then (depending on your local broadcast video standard) change the image size to

- 720x534 pixels (NTSC)
- 768x576 pixels (PAL)

Use the size that matches your local broadcast video standard (see Chapter 3 for more on video standards). Before you convert an image to one of these sizes, it should have an aspect ratio of 4:3. Common 4:3 image sizes include 800x600, 1024x768, and 1600x1200.

Again, you only need to convert the still image sizes if you plan to export the movie back to tape or DVD. If your movie will only be watched over the Web, you don't need to convert the image. The exact steps you should use to adjust the image size vary depending on the image-editing program you are using, but in most programs, they work something like this:

1. **Open the picture in your picture-editing program.**

   In the example shown here, I use Adobe Photoshop.

2. **Open the dialog box that controls the image size in your software.**

   In Adobe Photoshop, choose Image⇨Image Size.

3. **Remove the check mark next to the Constrain Proportions option.**

   Most editing programs have a similar constraining option (such as Constrain Size). Make sure it's disabled.

4. **Change the image size to 720x534 if you're working with NTSC video, or 768x576 if you're working with PAL video.**

   In Figure 12-1, I have changed the image size to 720x534 for NTSC video.

5. **Click OK and save the image as a JPEG image file.**

   Save the image using the JPEG format because it can then be used in just about any video-editing program. If your picture-editing software gives you a quality option when you save an image as a JPEG, choose the highest quality possible if the image will be used in a movie program. Other versatile formats include BMP, PICT, and TIFF, but you should check the help system for your video-editing program to see which formats are supported.

Although the image may look slightly distorted after you change the size, don't worry: It will look right when viewed with the rest of your video. Just remember that you only have to perform this modification if you plan to export the movie back to tape. If you plan to use your movie only on computer screens (or share it over the Internet), you don't need to change the aspect ratio of the image.

Figure 12-1: Use image-editing software to adjust the size of your picture.

Deselect this check box.

Getting the colors just right

In Chapter 3, I described how TV screens and computer monitors don't show colors exactly the same way. The biggest problem is that some colors in computer graphics simply cannot appear on a TV. These colors are often called *illegal* or *out-of-gamut* colors. Although this isn't a super-serious problem, you should still fix any out of gamut colors in your images before using them in video programs.

Unfortunately, the ability to easily fix out of gamut colors is usually only found in more advanced, expensive video-editing programs. Adobe Photoshop, which retails for about $600, is one such program. To fix the colors in an image, simply open the picture in Photoshop and choose Filter⇨Video⇨NTSC Colors. This filter removes all colors from the image that are out of gamut for NTSC TVs. A lower-cost option that also enables you to fix out-of-gamut colors is Photoshop Elements, which is available for a much more affordable price of about $100. You might even be able to get it for free with some digital cameras.

If your image-editing program doesn't have a video color filter, I wouldn't lose too much sleep over it. But if it does, fix those out-of-gamut colors so that the image looks better in the final video program.

Inserting Stills in Your Movie

Still images are pretty easy to use in your movies. Basically you import the image file, drop it into the timeline (just like a video clip), and then decide how long the still clip should remain on-screen. A lot of video programs use five seconds as a default time for still-image display, but you can easily change that yourself. The next few sections show you how to use stills in iMovie and Studio.

Using pictures in Apple iMovie

Apple's iMovie makes it easy to quickly use almost any picture on your computer in a movie project. Supported still-image formats include GIF, JPEG, PICT, TIFF, and BMP. The following sections show you how to import and use stills in your movies. I'll start by showing you how to use iPhoto — an important step because iMovie uses iPhoto for organizing stills.

Organizing photos with iPhoto

In the interest of simplification, iMovie uses Apple iPhoto for organizing still photos. This makes a lot of sense if you think about it; these two programs work together pretty seamlessly. If you already use iPhoto to organize your still images, you can move on to the next section. But if you haven't used iPhoto yet, you'll need to before you can use stills in iMovie.

I recommend that you have iPhoto version 2 (or later) installed on your computer. iPhoto comes preinstalled on any Mac that ships with OS X. If you don't have the latest version, visit `www.apple.com/iphoto/` to download this free program. Follow the instructions on the Apple Web site to download and install iPhoto. For more on using iPhoto, check out *iPhoto 2 For Dummies* by Curt Simmons, published by Wiley Publishing, Inc. When the program is installed, you can begin organizing your stills:

1. **Launch iPhoto by clicking its icon on the OS X Dock.**

   Alternatively, you can open it from the Applications folder on your hard disk.

2. **Choose File⇨Import.**

   The Import Photos dialog box appears, as shown in Figure 12-2.

Figure 12-2: Locate the picture file that you want to import into iPhoto.

3. **Browse to the folder that contains the picture or pictures that you want to import.**

   If you don't have a picture to import at this time, you can import the Chapter 12 sample photo `ch12_02.jpg` from the CD-ROM that accompanies this book.

4. **Select the picture and click Import.**

   To select multiple pictures, hold down the ⌘ key while clicking each picture file you want to import.

5. **The picture appears in the iPhoto library, as shown in Figure 12-2.**

   After the picture is imported into iPhoto, it's ready for use in iMovie as well.

6. **You can now quit iPhoto if you want by pressing ⌘+Q.**

Using images in your movie project

iMovie can use any picture that has been imported into iPhoto. In fact, importing the image into iPhoto first is mandatory — so if you haven't imported your still graphics into iPhoto (as described in the previous section), do so now. When that's done, you can start using still images in iMovie:

1. **Open iMovie and the movie project in which you want to use a still image.**

2. **Click the Photos button.**

   A collection of pictures that looks eerily similar to your iPhoto library will appear in the iMovie window as shown in Figure 12-3. If you've gone to the trouble of organizing photos into different albums in iPhoto, you can switch between those albums using the menu above the image browser.

3. **Click the clip that you want to use in the movie to select it.**

4. **Adjust the duration of the clip using the duration slider, which bears the images of a little tortoise and hare.**

   The default duration for a still clip is five seconds, but if you want it to display for less time, move the slider towards the bunny, er, I mean hare. If you want the clip to appear slow and steady, move the slider towards the tortoise.

   At about this time, you'll probably notice that your still graphic is moving in that little preview window in the upper-right corner of the screen. You'll also notice the words "Ken Burns Effect" just above. No, the Ken Burns Effect doesn't make Ken's smiling mug appear in your movie. It makes the camera appear to slowly zoom in or out on the still image, providing a visual continuity of movement that is otherwise broken by using a still image amongst moving video clips. (Ken Burns, the well-known documentary filmmaker responsible for such films as *The Civil War*, uses this trademark effect in a lot of his films.)

5. **Click the Start radio button near the top of the screen.**

   This allows you to adjust the zoom level at the beginning of the clip.

6. **Adjust the zoom slider.**

   If you don't want to zoom in on the image at all, drag the slider all the way to the left so that the zoom factor says 1.00.

7. **Click the Finish radio button near the top of the screen.**

   This allows you to adjust the zoom level at the end of the clip.

8. **Adjust the zoom slider again.**

   If you don't want to use a "Ken Burns" zooming effect, make sure that the zoom factor at the Start and Finish of the clip are the same.

9. **Click Preview to preview the results of the effect.**

   If you aren't happy with the effect, continue tweaking the zoom settings and previewing the results.

10. **When you're done making adjustments, drag the clip and drop it down on the timeline or storyboard to insert it in your movie.**

Figure 12-3: The iMovie Ken Burns Effect lets you add some apparent motion to your still pictures.

You can still adjust clips after they have been dropped into the timeline or storyboard. Select the clip, click the Photos button, and then adjust the duration or other attributes in the Photos panel. Click Apply when you're done making changes.

Using pictures in Pinnacle Studio

Studio is pretty conventional in the way it handles still graphics, which is a good thing. You can import pictures into the program and drop them in the timeline, just like almost every other video program available. Studio also makes it easy to use graphic images on top of video.

Placing images in the timeline

Studio's media album has a special section just for still graphics. You can open it by clicking the Show Photos and Frame Grabs tab as shown in Figure 12-4. By default, the album shows images in your My Pictures folder, but you can browse to a different folder if you wish by clicking the small folder icon shown in Figure 12-4. In this section, I'm using a sample JPEG picture that you can find in the Chapter 12 samples folder on the CD-ROM that accompanies this book.

To use a picture in your movie, simply drag it from the album and drop it on the video track of the timeline as shown in Figure 12-4. The default duration for a still graphic is four seconds. To change the duration, select the clip in the timeline and click the Open Video Toolbox button. The video toolbox window appears above the timeline. Type a new time into the Duration field at the top of the toolbox window. Alternatively, you can just click-and-drag on the edge of the still clip in the timeline to adjust its duration.

Making an overlay graphic

Pinnacle Studio treats still graphics as if they were titles. As explained in Chapter 9, you can put titles in either the video track or the title track of the timeline. If you put a title in the video track, it will be a full-screen title. If you put the title in the title track, it will be an *overlay* title, which means the words will appear over a video image.

Because Studio treats still photos like titles, you can use the title track to overlay your own custom graphics over a video image. To try it, follow these steps:

1. **Open Microsoft Paint by choosing Start⇨All Programs⇨Accessories⇨ Paint.**
2. **In Paint, choose Image⇨Attributes.**

   The Attributes dialog box opens.

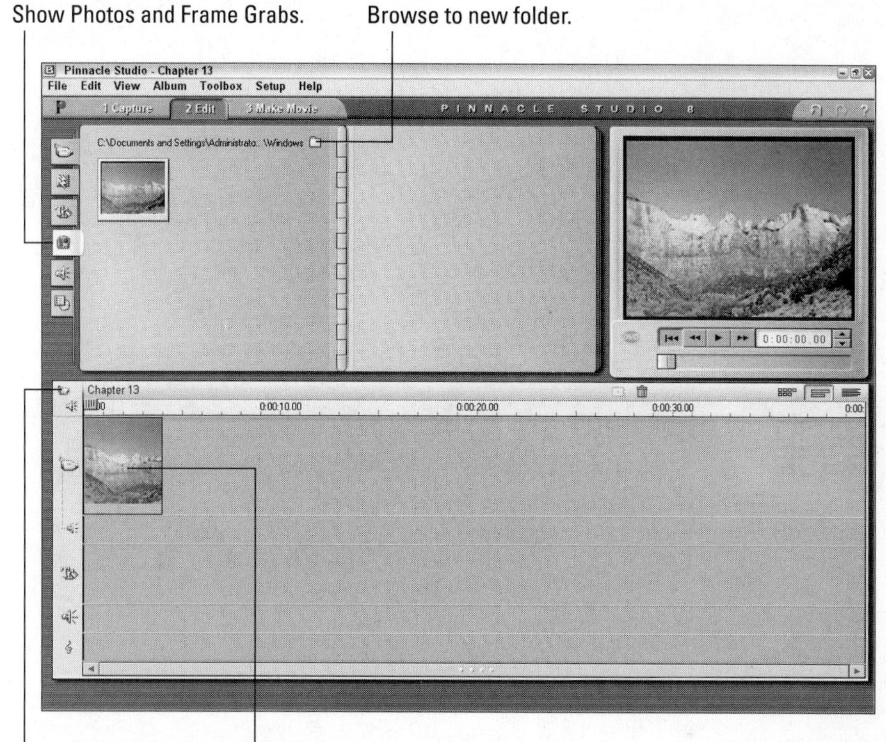

Figure 12-4: Still pictures are pretty easy to work with in Studio.

3. **Enter the height and width of the video image used in your video project, and then click OK.**

   If you're working with a DV-format project, make the image 720 pixels wide and 534 pixels high if you are working with NTSC footage, or 768 pixels wide and 576 pixels high if you are working with PAL footage.

4. **Draw an object.**

   Any old object will do, so let your creative juices flow. In Figure 12-5, you can see that I have drawn a UFO. No, really, that's what I was trying to draw.

5. **When you are done drawing, choose a color from the color palette that was not used in your UFO drawing (or whatever object you actually drew in Step 4).**

   My UFO drawing (Stop laughing! I mean it!) uses only gray, blue, and yellow, so I chose black from the palette.

6. **Click the Fill tool (see Figure 12-5).**

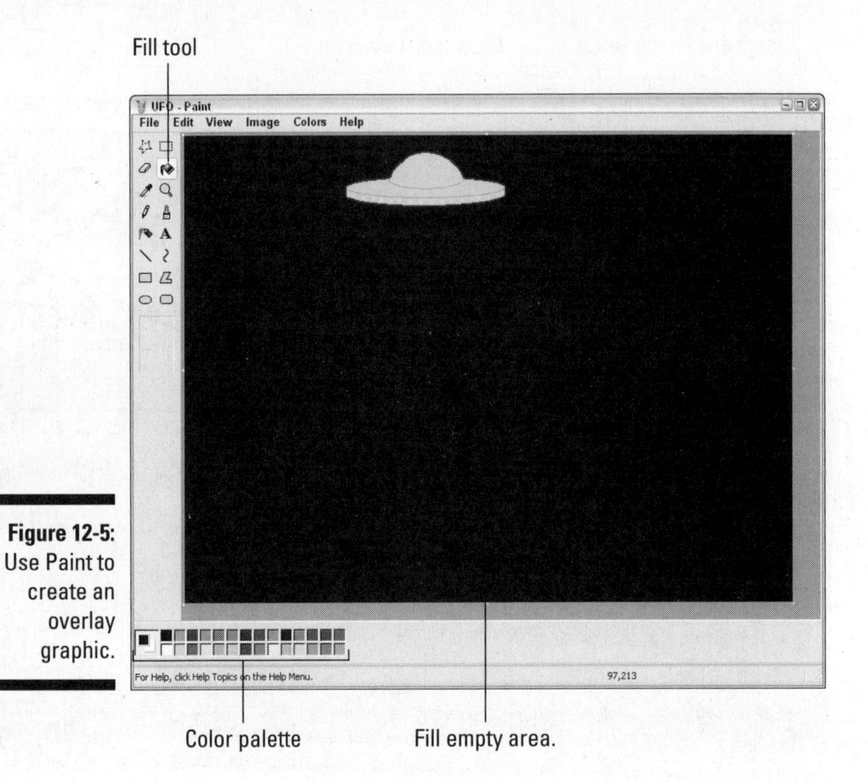

Figure 12-5: Use Paint to create an overlay graphic.

7. **Click a blank area of the image to fill it with the new color.**
8. **Save your picture and then close Paint.**
9. **Open Pinnacle Studio and the project in which you want to use the overlay graphic.**
10. **Click the Show Photos and Frame Grabs tab on the left side of the album.**
11. **If your drawing doesn't appear in the album, click the folder button at the top of the album and browse to the folder in which you saved your Paint graphic.**

    By default, Paint saves files in the My Pictures folder.
12. **Drag the overlay graphic from the album and drop it on the title track of the timeline.**

    The overlay image should now appear over the image in the video track, as shown in Figure 12-6.

Figure 12-6: You can really have some fun with overlay graphics.

You're probably wondering how Studio knows how to make the background of an overlay graphic transparent. When you place a graphic in the title track, Studio looks at the color of the pixel in the upper-left corner of the image and removes that color from the entire image. For example, if the upper-left pixel is black, all black pixels become transparent, whereas pixels that are blue, red, yellow, or other colors will remain. You may need to experiment with some different background colors for the best results. I have found that black works best most of the time.

Freezing Frames from Video

The first part of this chapter shows you how to use still pictures in your movies. But there will also be times when you might want to freeze a video image for posterity, so to speak. You may want to grab still pictures from video to help promote your movie, or you may want a still image of someone or something, and the video clip is the only image you have. Just keep in mind that the resolution of video images is really, really low compared to even the cheapest digital cameras, so frame grabs from video will have a lower quality. The lower quality will be especially apparent if you print the frame grab because image details will look blocky and pixelated. To freeze a frame of video and turn it into a still picture:

1. **Open the video clip that has the frame that you want to freeze.**

   If you don't have a video clip of your own, you can use the Chapter 12 sample video clip on the companion CD-ROM. In Pinnacle Studio, you must place the clip in the timeline before you can freeze a frame. To add a video clip to the timeline, simply drag it from the album to the timeline.

2. **Move the play head to the frame that you want to freeze.**

   In Figure 12-7, I am about to grab a frame that is 25 frames into the sample clip.

3. **Grab the frame.**

   In Apple iMovie, choose File➪Save Frame As. A Save As dialog box appears.

   In Pinnacle Studio, choose Toolbox➪Grab Video Frame. The Grab Video Frame toolbox appears, as shown in Figure 12-8. Choose the Movie radio button next to Grab From, click the Grab button, and then click Save to Disk.

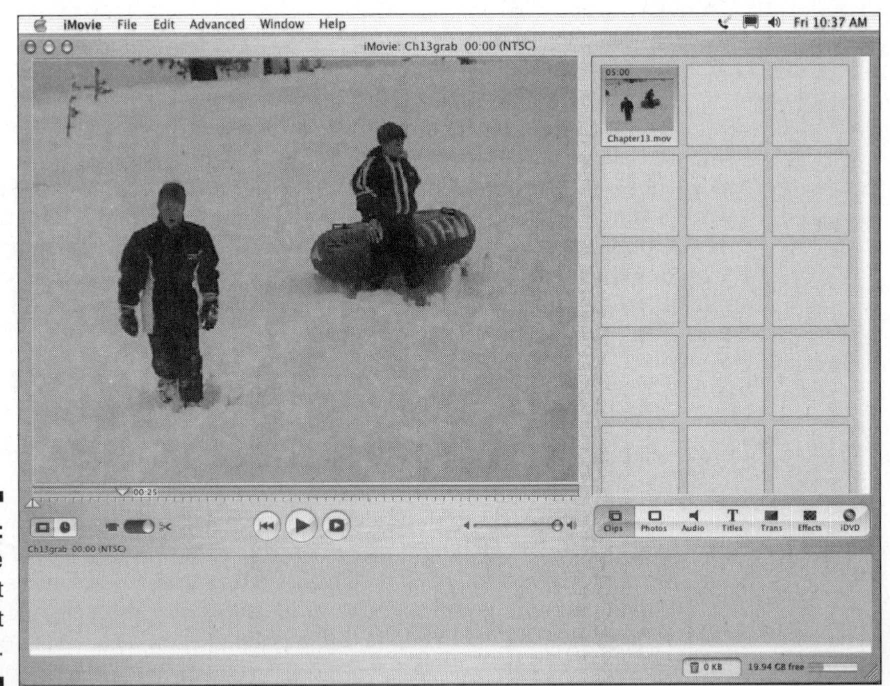

Figure 12-7: Locate the frame that you want to freeze.

4. **Save the frame.**

   In iMovie, name the picture and choose a format in the Format menu. The available formats are JPEG and PICT, but if you're not sure which format to choose, I recommend JPEG.

   In Studio, name the file and choose a format from the Save As Type menu. Several formats are available, but I recommend the JPEG format for the greatest versatility.

After you have grabbed a frame of video, you can use the image on a Web page to promote your movie, use it as a background image for a DVD menu (see Chapter 16 for more on making DVD menus), or share it as you would any other still photo.

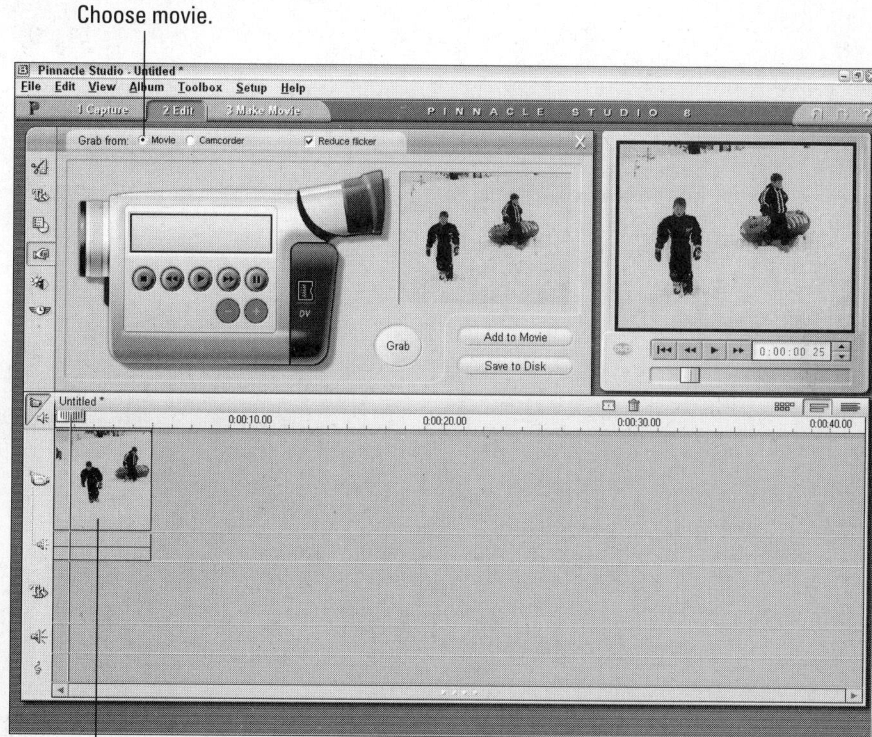

Figure 12-8: Use Studio's Grab Video Frame toolbox to freeze a frame of video.

Chapter 13

Wrapping Up Your Movie

In This Chapter

- Previewing your movie onscreen
- Previewing your movie on a TV
- Choosing a file format for export
- Exporting your movie

One thing I have learned about the movie-making process is that the most difficult part is finishing it. I always find a scene that should be shortened, or a title that could be styled differently, or an audio level that could be adjusted. You could practically spend an eternity making fine adjustments to your movie, even if it's only a couple of minutes long.

At some point, you just have to decide that the movie is pretty good and ready to be released. This chapter helps you preview your final project effectively and shows you how to begin the process of exporting your movie so that others can view it.

Previewing Your Movie

I could start and end this section by simply saying "Click Play" under the preview window in your editing program. If you want to see a bigger preview in Apple iMovie, click the Play Movie Fullscreen button shown in Figure 13-1. If you are using Pinnacle Studio, alas, you are limited to the tiny, inflexible size of the Studio preview window.

Of course, truly previewing a movie means a whole lot more than just clicking the Play button. To perform a truly effective preview, you must

- Draw a properly critical eye on the movie.
- Decide whether just previewing the movie on your computer screen is sufficient, or whether you need to preview it on a TV monitor.

Figure 13-1: Click the Play Movie Fullscreen button in iMovie for an even bigger preview.

The next two sections address both of these subjects. Warm up that coffee, sit back in your director's chair, and get ready to see what kind of movie magic you hath wrought.

You should also make sure that you preview your movie on the same type of equipment that your audience will use. This includes testing it in the player software that your audience will use if you plan to share your movie on the Internet or a recordable CD.

Casting a critical eye on your project

Of course, there's more to previewing your project than simply clicking Play. Consider carefully what you are actually previewing when you play your movie. Here are some ways to get the most out of previewing your project:

- **Watch the whole program from start to finish.** You may be tempted to periodically stop playback, reverse, and repeat sections, or perhaps even make tweaks to the project as you run it. This is fine, but to get a really good "feeling" for the flow of the movie, watch the whole thing start to finish — just as your audience will. Keep a notepad handy and jot down quick notes if you must.

- **Watch the program on an external TV.** If you plan to record your movie on videotape or DVD, previewing on an external monitor is very helpful. (See the next section in this chapter for a more detailed explanation.)
- **Have trusted third parties review the project.** Moviemakers and writers are often too close to their creations to be totally objective; an "outside" point of view can help a lot. Although I worked hard to write this book (for example), my work was reviewed by various editors and their feedback was invaluable. Movie projects benefit from a similar review process. Even if you want to maintain strict creative control over your project, feedback from people who were not involved with creating it can help you see it afresh.

Previewing on an external monitor

Your editing software has a preview window, and at first glance it probably seems to work well enough. But if you plan to record your movie on videotape or DVD, just previewing it on the computer screen can cause a couple of problems:

- **TVs usually provide a bigger view.** A larger TV screen reveals camera movements and other flaws that might not be obvious in the tiny preview window on your computer screen.
- **Computer monitors and TVs show color differently.** Colors that look right on your computer may not look the same on a TV.
- **Most TVs are interlaced.** As I described in Chapter 3, TVs are usually interlaced, whereas computer monitors are progressively scanned, or non-interlaced. Titles or other graphics that have very thin lines may look fine on your computer monitor, but when viewed on a TV, those thin lines may flicker or appear to shimmer or crawl.

In Chapter 2, I show you how to connect a TV to your computer for previewing video. The cheapest option is to connect your digital camcorder to your FireWire port, and then connect the TV to the analog output on your camcorder. If you have an analog capture card in your computer, such as the Pinnacle AV/DV board, you can connect your TV (and VCR) to the analog outputs for the card.

Apple iMovie makes previewing your work on an external TV monitor very easy. In iMovie, choose iMovie ⇨ Preferences. The iMovie Preferences dialog box appears as shown in Figure 13-2. Place a check mark next to Play Video

Through to Camera under Advanced options. Now that this option is enabled, whenever you play the timeline, the video will be sent out to your camcorder if it is connected to your FireWire port and turned on in Player or VTR mode. If you have a TV monitor hooked up to the camcorder's analog outputs, the video image should appear there as well.

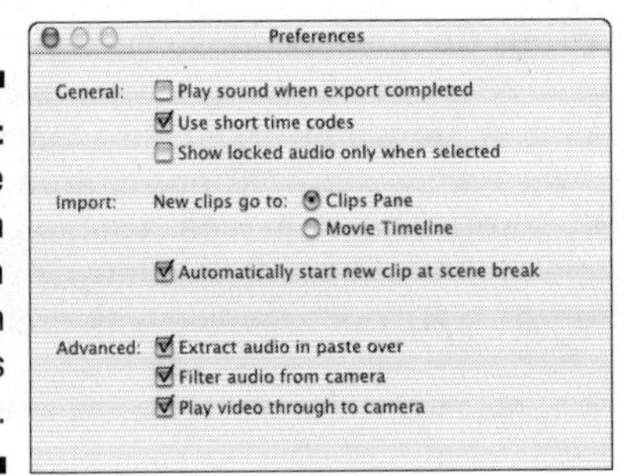

Figure 13-2: Enable the Play through to camera option in iMovie's preferences.

Pinnacle Studio also allows you to preview video on an external monitor, although it's a little more complicated. Basically, you have to export the movie as if you were done and ready to record it on tape. After the movie is exported for tape, you can play the export file and preview it on an external monitor as many times as you like without actually recording the tape. See Chapter 15 for steps on exporting movies for tape in Studio.

Exporting the Movie

After about the millionth preview, you'll probably come to the realization that your movie will never be perfect: If you don't just give up and release your movie, it will never be seen by anyone. *Star Wars* creator George Lucas once described this process as one of abandonment because most moviemakers almost never feel that a project is truly finished.

When you decide that it's time to "abandon" your project, you must export the movie so that others can view it. Before you export your movie, you must decide in which format you wish to export. After you've edited your movie in a program like Studio or iMovie, you aren't limited to showing your work to others on your computer. You have many options with which to share your movie:

- **Videotape:** Almost everyone you know probably has a VCR that plays VHS tapes. Sure, the quality and gee-whiz factor of VHS tapes isn't as high as DVD, but most of your audience really won't care. If you have an analog capture card such as a Pinnacle AV/DV board, you can connect a VCR directly to your computer and export your movie directly to a VHS tape. Otherwise export the movie back to a tape in your digital camcorder, and then dub the movie from the camcorder tape to a VCR. (Chapter 15 explains the process of exporting to tape in greater detail.)

- **DVD:** The DVD (Digital Versatile Disc) player is quickly supplanting the VCR as the home video player of choice for many people. Tens of millions of households already have DVD players in their home entertainment systems, and many modern computers can play DVD movies as well. DVD recorders are now relatively affordable, so making your own DVDs is pretty easy. See Chapter 16 for details.
- **VCD/S-VCD:** If you don't have a DVD burner yet, you can make DVD-like movie discs called VCDs (Video Compact Discs) or S-VCDs (Super VCDs) using a regular CD-R drive and blank CDs. VCDs can hold about one hour of video, and S-VCDs hold 20 minutes of higher quality video. VCDs and S-VCDs can be played in most DVD players. Chapter 16 also shows you how to make VCDs and S-VCDs.
- **Internet:** The Internet is a popular place to share video, but remember that video files are usually very big and most people still have relatively slow Internet connections. If you plan to export your movie for use on the Internet, use a Web-friendly format such as Apple QuickTime, RealNetworks RealVideo, or Windows Media Video. These formats usually have fairly low quality, but the trade-off is much smaller file sizes. When choosing one of these formats, make sure that your intended audience has software that can play your movies.
- **Video file:** Some programs allow you to simply save the movie as a file. Common video file formats include AVI and MPEG. These files can be stored on your computer, or burned onto a CD so that others can watch your movie on their computers as well. Just remember that the AVI and MPEG formats are generally too big for anyone but broadband users to access online. AVI or MPEG files are useful, however, if you plan to edit the movie further using another editing program.

If you're using Apple iMovie, QuickTime is the only file format available for export. In Chapter 14, I show you how to export your movie in QuickTime format. But if you're using Pinnacle Studio, you have more options. To export your movie to an AVI or MPEG file:

1. **In Studio, choose View⇨Make Movie.**

   The Make Movie screen appears, as shown in Figure 13-3.

2. **Click either AVI or MPEG on the left side of the Make Movie window.**

   Choose the format to which you want to export. Generally speaking, AVI (Audio Video Interleave) files are a little bigger than MPEG (Moving Picture Experts Group) files. Anyone with a semi-modern Windows PC should be able to view either format. Another benefit of choosing MPEG is that Macs can usually open them, whereas AVI files are usually not Mac-compatible.

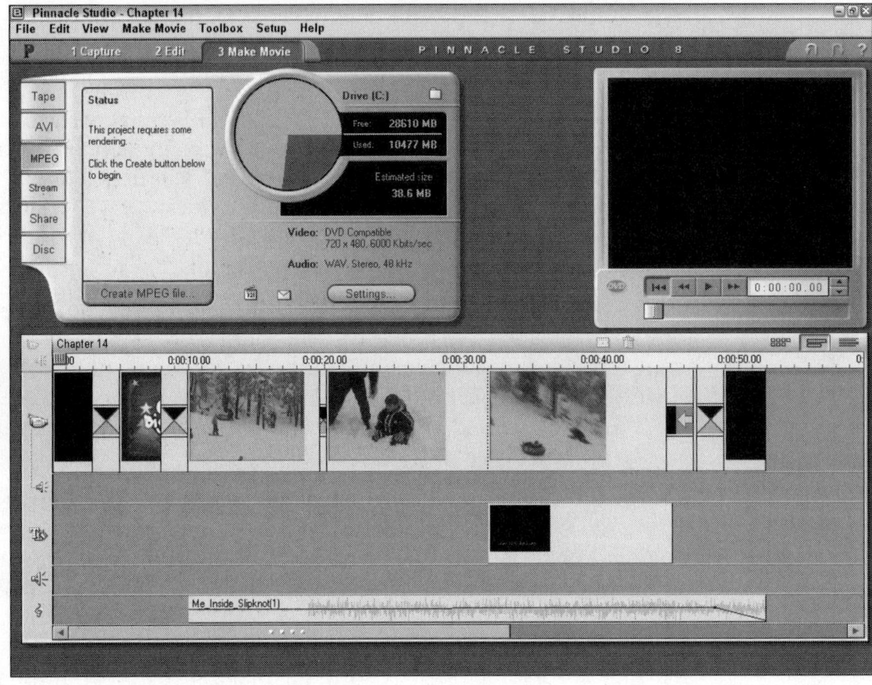

Figure 13-3: Studio's Make Movie view is where you export your movie to the rest of the world.

3. **Click the Settings button.**

   The Pinnacle Studio Setup Options dialog box appears, as shown in Figure 13-4. If you chose AVI in Step 2, the Make AVI File tab will be in front. If you chose MPEG in Step 2, the Make MPEG File tab will be in front.

4. **Adjust settings and click OK.**

   I'll describe the various AVI and MPEG export settings in the following two sections.

5. **Review the estimated file size above the Settings button in the Make Movie window.**

   The estimated file size is just that — an estimate. It's usually pretty close to what the final file size will be, but seldom exact. If the file seems too big, adjust settings so that the video picture is smaller or the audio quality is lower.

6. **Click Create AVI File or Create MPEG file to create your movie.**

7. **In the dialog box that appears, choose a folder in which to save your movie, and give it a filename.**

8. **Click OK.**

   The export process will probably take a few seconds or even minutes, depending on the length of your movie and the number of effects.

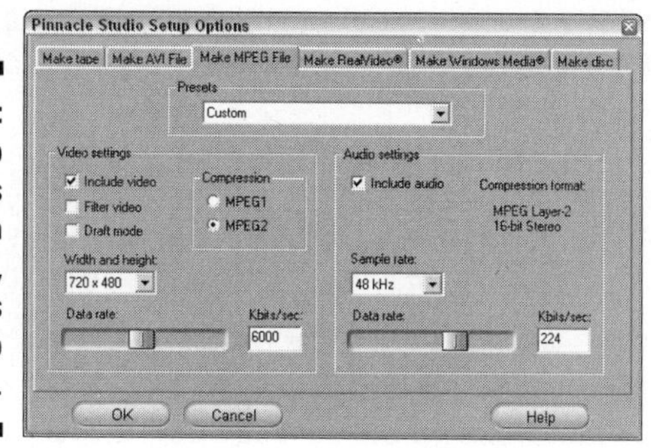

Figure 13-4: The Setup Options dialog box in Pinnacle, with its MPEG tab visible.

One reason you might want to export a movie as an AVI or MPEG file is if you plan to edit it using other software. For example, if I want to export a movie for the Web in Windows Media Video (WMV) format, I prefer to use the export feature in Windows Movie Maker. When I'm done editing in Studio, I export the movie in AVI format, which can be easily imported into Movie Maker. After the AVI file is imported into Windows Movie Maker, I use that program to export a WMV file.

AVI is a good format to use if you want to export video for many Windows-based video-editing programs, including Windows Movie Maker or Adobe Premiere for Windows, but you may find that some editing programs prefer MPEG. Pinnacle's Hollywood FX Pro, for example, works better with MPEG-format video. And of course, if you plan to use the exported file on a Mac, you'll find that the MPEG format will be much more compatible with Mac multimedia software.

If you aren't sure whether you have a specific need for one format or the other, it probably doesn't matter very much which one you choose. AVI files are usually a little bigger than MPEG files, but as I said earlier, if you're concerned about file sizes — say, for example, you plan to share the movie on the Internet — you really ought to be exporting the movie in a Web-friendly format like RealVideo or Windows Media. See Chapter 14 for more on exporting in either of those formats.

Choosing AVI settings

The AVI format was developed several eons ago by Microsoft as the file format of Video for Windows video files. AVI files can use one of several codecs to compress video. Codecs control exactly how the audio and video is compressed. Check out the sidebar "Decoding codecs" for a more detailed description of codecs. As you can see in Figure 13-5, Studio's Setup Options dialog box makes it pretty easy to choose a codec or modify other settings. The video settings include

- **Include Video in AVI File:** I'm guessing you'll want to leave this option checked for 99.9975% of all your projects. If you uncheck this option, only audio will be included in the file.
- **List All Codecs:** I recommend you leave this option unchecked. Many multimedia programs install codecs on your computer, and if you check the "List all codecs" option, each and every codec installed on your PC will appear in the list, even ones that aren't compatible with Studio.
- **Options:** Some codecs have further options you can adjust. I generally recommend that you don't mess with these options.
- **Compression:** This menu allows you to choose a codec. Many multimedia programs install codecs on your computer, so the list of codecs may vary depending on what is installed on your computer. The Cinepak codec (which comes with Studio) is pretty good for all around use, and is particularly recommended for movies that will be recorded onto a CD.
- **Width and Height:** These indicate the size of your video image in pixels. Full-size DV-format video is 720 by 480 pixels. Reducing the size of the picture can greatly reduce the file size, but make sure you keep the ratio between the height and width the same, or your video image may look distorted.
- **Frames/Second:** This controls the frame rate of the video file. Full-quality DV-format video uses a frame rate of 29.97 frames per second (fps). Reducing the frame rate decreases playback quality, but it also greatly reduces file size.
- **Quality or Data Rate:** This slider control isn't available with all codecs, but when it is available, it allows you to adjust the quality up or down to control file size. A smaller data rate greatly reduces file size.

The Make AVI File tab of the Setup Options dialog box also contains audio settings. These settings include

- **Include Audio in AVI File:** If you only want to export a video picture, uncheck this option.
- **List All Codecs:** As with video codecs, I recommend you leave this option unchecked.
- **Type:** This menu allows you to pick a compression method for the audio. The default setting is usually best.
- **Channels:** Here you can choose 8-bit or 16-bit mono or stereo audio. As with other settings, higher quality means bigger files.
- **Sample Rate:** CD-quality audio uses a sample rate of 44.1 kHz (kilohertz), but most digital camcorders record at 48 kHz. Higher sample rates mean higher quality, and of course bigger files.

Decoding codecs

Digital video contains a lot of data. If you were to copy uncompressed digital video onto your hard disk, it would consume 20 MB (megabytes) for every second of video. Simple arithmetic tells us that one minute of uncompressed video would use over 1GB. Even with a 60GB hard drive, you would have room for only about 50 minutes of uncompressed video, assuming that big drive was empty to begin with. Dire though this may seem, storage isn't even the biggest problem with uncompressed video. Typical hard drive busses and other components in your computer simply can't handle a transfer rate of 20MB per second, meaning that some video frames will be dropped from the video. Dropped frames are described in greater detail in Chapter 5.

To deal with the massive bandwidth requirements of video, digital video is compressed using compression schemes called *codecs* (*co*mpressor/*dec*ompressor). The DV codec, which is used by most digital camcorders, compresses video down to 3.6MB per second. This data rate is far more manageable than uncompressed video, and most modern computer hardware can handle it without trouble. When you capture DV video from a camcorder using a FireWire interface, a minute of video consumes just over 200MB of hard disk space. Again, most modern computers can manage that.

Why do codecs matter to you? When you choose a file format for exporting your movie, you're also usually choosing a codec to compress your movie (whether you realize it or not). Usually your export software automatically chooses a codec for you, but as you've seen in this chapter, you can also usually choose a specific codec if you wish. Some codecs compress video more than others. Generally speaking, the more video is compressed, the more quality you lose. In most cases I recommend you use the default codec chosen by Studio (or whatever program you are using) when you select an export format.

My very favorite button on the whole Make AVI File settings tab is the Same As Project button. As you goof around with settings, it's easy to forget what your original settings were. If you want to make sure that the movie you export has the same quality as your source footage, click this button. All of the settings will be automatically set to match your project settings.

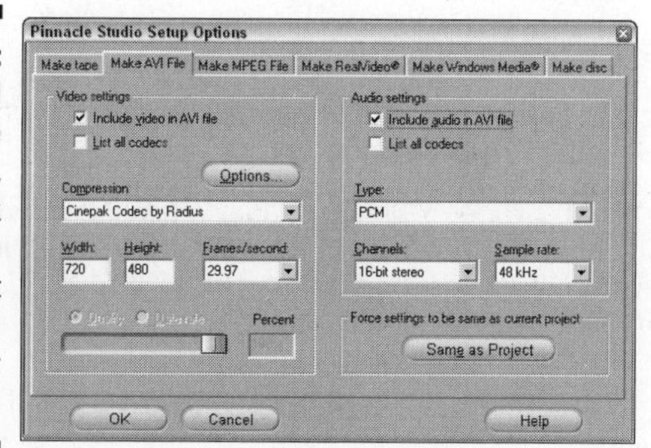

Figure 13-5: If the AVI settings are confusing, just click that Same as Project button on the lower right.

Setting MPEG settings

I like the MPEG format because it is easy to use and — most importantly — widely supported among Mac and PC users. MPEG is actually a family of multimedia file standards. There are currently four MPEG standards:

- **MPEG-1:** This is the oldest version of the standard. A drawback of MPEG-1 is that it has a maximum picture size of 352 by 240 pixels.
- **MPEG-2:** This standard offers much higher video quality and full-size video. In fact, MPEG-2 is the format used by DVD movies, so if you burn your movies onto DVD, this is the format you'll use. (I cover burning your own DVDs in Chapter 16.) MPEG-2 files require special MPEG-2 player software, such as a DVD player program.
- **MPEG-3:** Usually abbreviated MP3, this file format only contains audio and is a popular file format for music files today.
- **MPEG-4:** This newer standard is a cross of MPEG video and Apple QuickTime to produce video with very, very small file sizes. Support for this format is still limited.

As you can see in Figure 13-4, Studio's MPEG settings dialog box is a lot simpler than the AVI settings. Review the following settings as you get ready to export your movie in MPEG format:

- **Presets:** Studio provides a selection of presets based on how the file will be used. For example, if you plan to record the movie on a VCD (see Chapter 16 for more on this format), choose the VideoCD preset from the Presets menu. If possible, I recommend that you stick with one of the presets, but if you absolutely must fiddle with the rest of the MPEG settings, choose Custom from the Presets menu. If you use one of the presets, you can simply click OK to close the Setup Options dialog box. Additional settings can only be adjusted if you choose Custom.
- **Include video:** You'll want to leave this option checked unless you only want to generate an audio file.
- **Filter video:** If you're working with a smaller frame size, check this option to smooth the appearance of the video image.
- **Draft mode:** Check this option if you just want to quickly produce a low-quality file for previewing purposes.
- **Compression:** Choose MPEG1 or MPEG2. Again, MPEG-2 can provide higher quality, but it requires special player software. MPEG-2 is used on movie DVDs.

- **Width and height:** Select a frame size for your video image here. Smaller frame sizes mean smaller files.
- **Data rate:** Most of the time you can leave these sliders alone, but you can use them to fine-tune the quality and file size. The slider on the left controls video data rate, and the slider on the right controls audio data rate.

The Make MPEG File tab of the Setup Options dialog box also includes a couple of audio settings. Audio options are

- **Include audio:** Uncheck this option if you only want to output video.
- **Sample rate:** 44.1 kHz is CD quality, but most digital camcorders can record at 48 kHz. The higher the sample rate, the higher the audio quality (and file size).
- **Data Rate:** Use this slider to fine-tune the quality of the audio. Slide it left to reduce quality and file size, or slide it right to increase quality and file size. As you adjust the slider, you see the number in the Kbits/sec box change. Generally I recommend leaving the Data Rate slider alone unless you really need to squeeze a couple more kilobytes of file size out of your movie file.

Part IV
Sharing Your Video

"Do you think the 'Hidden Rhino' clip should come before or after the 'Waving Hello' video clip?"

In this part...

When you're done editing your movie, it's time to share your work with others. Thanks to digital video, getting your movies out into the world has never been easier. This part shows you how to share your movies online, on videotapes, or even on DVDs that you record yourself.

Chapter 14

Putting Your Movies on the Internet

In This Chapter

- Choosing a video format
- Exporting your movie
- Optimizing your movie's export settings for the Internet
- Finding an online home for your movies
- Putting your movies on the Web

Since the first home movie cameras became popular about 50 years ago, friends, family, and neighbors have taken part in a vaunted tradition: Falling asleep in front of someone else's home movies. Until recently, this tradition required that you, the moviemaker, lure your unsuspecting audience over to your house, baiting them with the promise of good food and drink. But now thanks to the Internet, you can bore, er, I mean, *entertain* many more people with your movies without ever having to prepare a single hors d'oeuvre tray.

Software companies have tried to make the process of putting your movies online easier and easier. Nearly all video-editing programs — including Apple iMovie, Pinnacle Studio, and Windows Movie Maker — provide tools to help make your movies ready for sharing online.

Easily the most important consideration when you want to share movies on the Internet is file size. Movie files are usually really big, and (believe it or not) a majority of Internet users still have slow dial-up modem connections. This means that big movie files will take a long time to download, and the long download time might even discourage your audience members from trying in the first place.

This chapter helps you prepare your movies for online use. I'll show you how to use the most efficient file formats so that your video files aren't too big, and I'll show you how to find an online home for your movies once they're ready to share.

Choosing a Video Format

Many different video formats are available for the movies you edit on your computer. Each format uses a different codec. (I explain codecs in greater detail in Chapter 13, but a codec, short for *compressor/decompressor,* is a software tool used for making multimedia files smaller.) Common video file formats include MPEG and AVI, but these two formats are usually not suitable for movies you plan to share online because they have big file sizes. Three other popular formats, however, are perfectly suited to the online world:

- **QuickTime (.QT):** Many Windows users and virtually all Macintosh users have the QuickTime Player program from Apple. QuickTime is the only export format available with iMovie. Pinnacle Studio cannot export QuickTime movies, but some more advanced Windows programs like Adobe Premiere can.
- **RealMedia (.RM):** This is the format used by the popular RealPlayer, available for Windows and Macintosh systems, among others. Pinnacle Studio can export RealMedia-format video.
- **Windows Movie Video:** This format requires Windows Media Player. Almost all Windows users and some Macintosh users already have it. Both Pinnacle Studio and Windows Movie Maker can export Windows Media Video.

Each of these three video formats has strengths and weaknesses. Ultimately the format you choose will probably depend mainly on the editing software you're using — for example, if you're using iMovie on a Mac, QuickTime is your only option. Later in this chapter, I introduce you to the various software programs that are used to play these Web-friendly video formats.

Streaming your video

Doing stuff on the Internet usually means downloading files. For example, when you visit a Web page, files containing all the text and pictures on that Web page are first downloaded to your computer, and then your Web-browser program opens them. Likewise, if someone e-mails you a picture or a document with yucky work stuff, your e-mail program actually downloads a file before you open it.

Downloading files takes time, especially if they're big video files. You sit there and you wait. And wait. And wait. Finally the movie file is done downloading and starts to play, but by then you've left the room for a cup of coffee again. But software designers are crafty folk, and they've devised methods of getting around the problem of waiting a long time for downloads. They've come up with two basic solutions:

- **Streaming media:** Rather than downloading a file to your hard drive, streaming files can be played as the data streams through your modem. It works kind of like a radio, where "data" streams through in the form of radio waves, and that data is immediately played through the radio's speakers as it is received.

  With streaming audio or video, no file is ever saved on your hard drive. To truly stream your movies to other people, your movie files need to be on a special *streaming server* on the Web. There is a remote possibility that your Internet service provider offers a streaming media server, but most service providers do not.

- **Progressive download:** Newer video-player programs can "fake" streaming pretty effectively. Rather than receiving a movie signal broadcast over the Internet like a radio wave, viewers simply click a link to open the movie as if they were downloading the file. In fact, they *are* downloading the file — but as soon as enough of the file has been received, the player program can start to play. The program doesn't need to wait for the whole file to download before it starts. Current versions of QuickTime, RealPlayer, and Windows Media Player all support progressive download.

  The really cool thing about progressive download is that you don't need any special kind of server to host the files. Just upload the video file to any server that has enough room to fit it in.

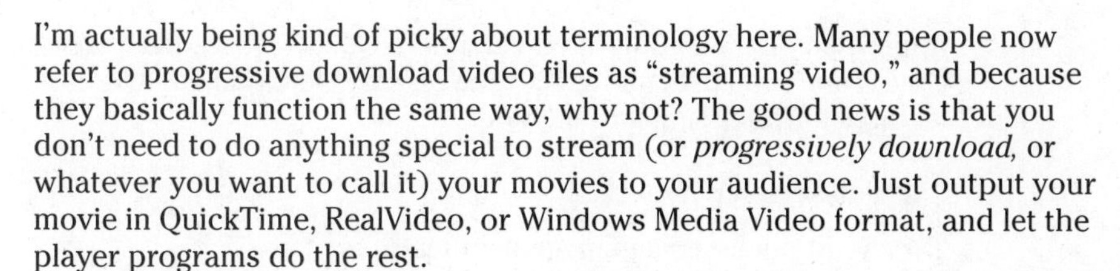

I'm actually being kind of picky about terminology here. Many people now refer to progressive download video files as "streaming video," and because they basically function the same way, why not? The good news is that you don't need to do anything special to stream (or *progressively download,* or whatever you want to call it) your movies to your audience. Just output your movie in QuickTime, RealVideo, or Windows Media Video format, and let the player programs do the rest.

Comparing player programs

When you share your movie over the Internet, you're actually sharing a file. Make sure that your intended audience can open that file. Different movie-file formats require different programs for playback. The following sections introduce you to the three most common programs used for playing movies from the Internet.

QuickTime

Apple QuickTime (see Figure 14-1) is perhaps the most ubiquitous media player in the personal computer world today, which makes it a good overall choice for your audience. QuickTime is available for Macintosh and Windows systems and is included with Mac OS 9 and higher. QuickTime can play MPEG and QuickTime media. The QuickTime Player also supports progressive download, where files begin playing as soon as enough has been downloaded to allow continuous playback. The free QuickTime Player is available for download at

```
www.apple.com/quicktime/download/
```

Figure 14-1: Apple's QuickTime is one of the most common and best media players available.

Apple also offers an upgraded version of QuickTime called QuickTime Pro. QuickTime Pro costs about $30 (the regular QuickTime Player is free). Key features of QuickTime Pro include

- Full-screen playback
- Additional media management features
- Simple audio and video creation and export tools
- Advanced import/export options

If you already have iMovie (and therefore regular QuickTime), you don't absolutely need the extra features of QuickTime Pro. Your audience really doesn't need QuickTime Pro either (unless of course they want to watch movies in full screen). The standard QuickTime Player should suffice in most cases. Apple iMovie exports QuickTime-format files. If you're a Windows user, QuickTime Pro allows you to convert MPEG files to QuickTime format. Some advanced Windows editing programs (such as Adobe Premiere) can also export files in QuickTime format.

RealPlayer

Another very popular media player is RealPlayer from RealNetworks. RealPlayer is available for Macintosh, Windows, and even Unix-based systems. The free RealPlayer software is most often used for RealMedia streaming media over the Internet, though it can also play MPEG-format media as well. Pinnacle Studio allows you to export movies in the RealMedia format using the "Streaming" option in the Make Movie window. I'll show you how to export a RealMedia movie later in this chapter. To download the RealPlayer in its various incarnations, visit

```
www.real.com/
```

Although RealNetworks does offer a free version of the RealPlayer (pictured in Figure 14-2), you have to look at their Web site carefully for the "Free RealOne Player" link before you can download it. RealNetworks offers other programs as well — and though they're not free, they offer additional features. RealNetworks has specialized in the delivery of streaming content, and they offer a variety of delivery options. You can use their software to run your own RealMedia streaming server, or you can outsource such "broadcast" duties to RealNetworks.

Figure 14-2: RealPlayer is a very popular media player, often used for streaming media on the Internet.

A complaint often heard about RealPlayer is that the software tends to be intrusive and resource-hungry once installed — and that the program itself collects information about your media-usage habits and sends that information to RealNetworks. Although RealPlayer is extremely popular, consider that some folks out there simply refuse to install RealNetworks software on their computers. RealMedia is an excellent format, but I recommend that you offer your audience a choice of formats if you plan to use it; include (for example) QuickTime or Windows Media Video.

Windows Media Player

Microsoft's Windows Media Player (version 7 or newer) can play many common media formats. I like to abbreviate the program's name *WMP* because, well, it's easier to type than Windows Media Player. WMP comes pre-installed on computers that run Windows Me or Windows XP. Although the name says "Windows," versions of WMP are also available for Macintosh computers that run OS 8 or higher. Figure 14-3 shows Windows Media Player running in OS X. WMP is even available for Pocket PCs and countless other devices! WMP is available for free download at

```
www.microsoft.com/windows/windowsmedia/download/
```

Windows Media Player can play video in MPEG and AVI formats. Although Pinnacle Studio can output both of these formats, they're not terribly useful for online applications because they create big files and have an appetite for resources. Windows Media Player can also play Windows Media Video (WMV) format, and Studio can output that as well (by using the Streaming option in the Make Movie window, I'll show you how later in this chapter). I like the WMV format because it provides decent quality (for Web movies) with remarkably small file sizes.

Figure 14-3: Windows Media Player is required for viewing Windows Media-format movies.

What are the compelling reasons for choosing WMP (the Macintosh version is shown in Figure 14-3) over other players? Choose Windows Media Player as your format if

- **Most or all of your audience members use Windows.** Most Windows users already have WMP installed on their systems, so they won't have to download or install new software before viewing your Windows Media-format movie.

- **You want the look, but not the expense and complexity, of streaming media.** If you don't want to deal with the hassle of setting up and maintaining a streaming-media server, Windows Media format files can provide a workable compromise. WMP does a decent simulation of streaming media with *progressive downloadable video*: When downloading files, WMP begins playing the movie as soon as enough of it is downloaded to ensure uninterrupted playback.
- **You're distributing your movie online and extremely small file size is more important than quality.** The Windows Media format can offer some very small file sizes, which is good if your audience will be downloading your movie over slow dial-up Internet connections. I recently placed a 3:23-long movie online in Windows Media format and the file size was only 5.5MB (megabytes). Of course, the movie was not broadcast quality, but because most of my friends and family still have slow dial-up modem connections to the Internet, they appreciated the relatively small download size.

Exporting Movies for the Online World

When you output a movie from Pinnacle Studio in MPEG or AVI format (as described in Chapter 13), you run into a familiar trade-off: Although those two formats can offer high quality, they're usually too big to use on the Internet, whether you plan to place your movie on a Web page or e-mail it to some friends. For online use, the QuickTime, RealMedia, and Windows Media Video formats are much better. The following sections show you how to export Web-friendly movies from Apple iMovie and Pinnacle Studio.

Making QuickTime movies with iMovie

If you're using Apple iMovie and you want to make movies in QuickTime format, you're in luck. QuickTime is the only movie file format that iMovie can produce. QuickTime movies can be played using the QuickTime Player program, which is available for free for Windows and Macintosh systems. If you want to output in a different format, such as RealMedia, you'll need to use more advanced software such as Final Cut Express.

Pinnacle Studio can export movies in RealMedia or Windows Media Video format, but not QuickTime. To create QuickTime movies in Windows, you'll need QuickTime Pro, or a more advanced editing program such as Adobe Premiere.

E-mailing your movies

The World Wide Web seems to get all the attention these days, but I think that e-mail, more than anything else, revolutionized the way we communicated during the last decade. Most of your friends, relatives, and business associates probably have e-mail addresses, and you probably exchange e-mail messages with those folks on a regular basis.

E-mail is already a great way to quickly share stories and pictures with others, and now that you're making your own movies, it only seems natural to start e-mailing your movie projects to friends as well. Before you do, keep in mind that movie files tend to be really big. Most e-mail accounts have file-size limitations for e-mail attachments, sometimes as low as 2MB. Other e-mail accounts don't allow any file attachments at all. And of course, many people still have slow dial-up modem connections to the Internet, meaning it will take them a long time to download a movie you send them.

If you want to e-mail a movie to someone, first ask the person whether it's okay to do so. Send an initial e-mail that says something like, "Hi there! I just finished a really awesome movie and I want to send it to you. The movie is in QuickTime format and the file is 1.3 MB. Can I e-mail it to you?" Most people will probably say yes, and they'll appreciate that you took the time to ask.

Exporting a QuickTime movie from iMovie is pretty simple. The QuickTime format offers a variety of quality and output settings that you can adjust, and iMovie provides several easy-to-use presets. You can also customize export settings if you wish. To export a QuickTime movie:

1. **When you're done editing your movie in iMovie, choose File⇨Export.**

   The iMovie: Export dialog box appears, as shown in Figure 14-4.

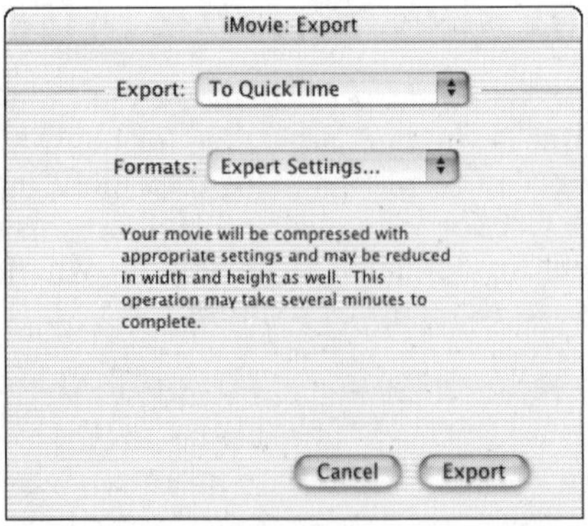

Figure 14-4: Choose a QuickTime preset from the Formats menu, or choose Expert Settings.

2. **Choose To QuickTime from the Export menu.**

3. **Choose the best preset for the way you plan to distribute your movie from the Formats menu.**

   iMovie provides three preset formats for export: Email, Web, and Web Streaming. Table 14-1 lists the specifications for each preset, as well as the CD-ROM and Full Quality DV presets. Unless your movie is very short, the CD-ROM and Full Quality DV presets generally produce files that are too big for online use. If you want to fine-tune your own settings, choose Expert Settings from the Formats menu.

4. **Click Export.**

   A Save Exported File As dialog box appears.

5. **If you chose a preset format, give your movie a filename, choose a folder in which to save it, and click Save to save your movie to a file and finish the export process.**

   If you chose a preset format in Step 3, you're done! But if you choose Expert Settings in the Formats menu, you still have a few more steps to complete in the export process. Your Save Exported File As dialog box will look like Figure 14-5 and you will have some additional settings to adjust. At the bottom of the dialog box, you can choose presets from the Use menu, or click Options. If you click Options, the Movie Settings dialog box appears.

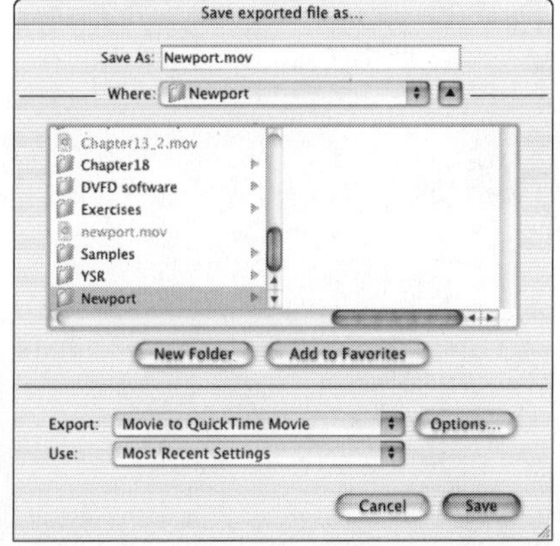

Figure 14-5: Name and save your movie file.

6. **(Optional) In the Movie Settings dialog box, leave the check marks next to Video and Sound if you want to include both in your movie.**

7. **(Optional) In the Movie Settings dialog box, click the Settings button under Video in the Movie Settings dialog box and adjust video settings.**

The Compression Settings dialog box appears, as shown in Figure 14-6. Start by choosing a codec from the menu at the top of the dialog box. The Sorenson or H.263 codecs are pretty good for most movies, and the Motion JPEG A codec works well for movies that will be played on older computers. MPEG-4 — which I've chosen in Figure 14-6 — provides a superior balance of quality and compression, but viewers must have QuickTime 6 (or later) to play it.

Adjust the Quality slider and preview how your video image will be affected. The Best quality setting provides better picture quality, but also increases the file size.

In the Frames Per Second menu, you can choose a frame rate (in Figure 14-6, I've chosen 15 frames per second), or just choose Best from the menu to let iMovie automatically determine a good frame rate. More frames per second increase file size.

I recommend that you leave the Key Frame Every *x* setting alone. (The default value is 24 — that is, a key frame occurs once every 24 frames — and a smaller number means *more* key frames.) Key frames help QuickTime compress and decompress the movie. More key frames provide better quality, but they also increase file size.

Click OK when you're done adjusting Compression Settings.

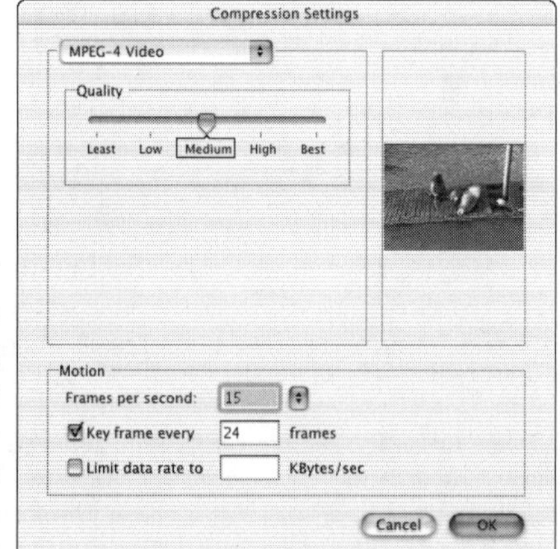

Figure 14-6: Adjust video compression settings here.

8. **(Optional) In the Movie Settings dialog box, click the Filter button.**

   The Choose Video Filter dialog box appears. Here you can apply filters to your video image that can blur, sharpen, recolor, brighten, or perform a variety of other changes to the picture. Most filters also have adjustments that you can make using slider controls. A preview of your video image appears in the Choose Video Filter dialog box so that you can see the affects of the various filters. I don't usually find these filters very useful because normally I've already applied filters or effects to my video during the editing process. After making your selections, click OK to close the Choose Video Filter dialog box.

9. **(Optional) In the Movie Settings dialog box, click the Size button.**

   The Export Size Settings dialog box appears. Choose either Use Current Size, or choose Use Custom Size and enter a custom width and height in pixels. As shown in Table 14-1, a common frame size for Web movies is 240 pixels wide by 180 pixels high. Click OK to close the Export Size Settings dialog box.

10. **(Optional) In the Movie Settings dialog box, click the Settings button under Sound.**

    The Sound Settings dialog box appears as shown in Figure 14-7. For online movies, choose QDesign Music 2 in the Compressor menu. Reduce the sampling rate using the Rate box (for online use I recommend 22.050 kHz or lower). Switching from Stereo to Mono will also reduce the file size. Click OK to close the Sound Settings dialog box.

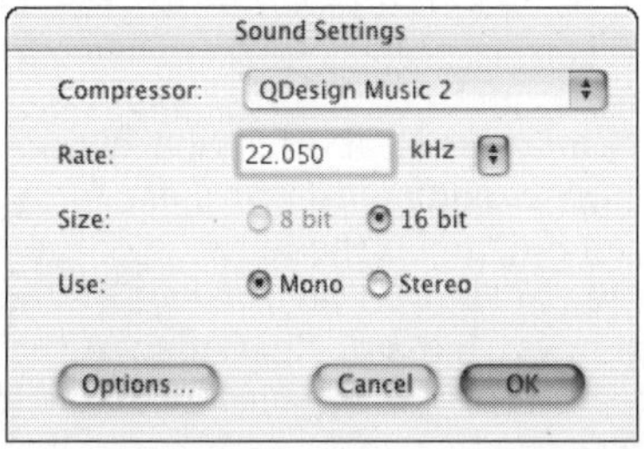

Figure 14-7: Make sound settings here.

11. **(Optional) In the Movie Settings dialog box, leave the Prepare for Internet Streaming option checked if you want to take advantage of streaming or progressive download for this movie, and choose the Fast Start option in the Streaming menu.**

12. **Click OK to close the Movie Settings dialog box, and then click Save in the Save Exported File As dialog box.**

    The movie will be exported using the settings you provided.

When the export process is complete, preview your movie in QuickTime, and check the file size of the movie. If the movie file is too big, re-export it using the lower quality settings (such as a smaller frame size, lower frame rate, or lower sample rate for the audio). If the movie is smaller than you expected, you may want to re-export it using slightly higher quality settings. When you're done, you can share your QuickTime movie file online by attaching it to an e-mail or placing it on a Web site. (Later in this chapter, I show you how to find and establish an online screening room for your movie.)

Table 14-1 iMovie 3 QuickTime Export Presets

Preset	*Frame Size (pixels)*	*Frame Rate (frames per second)*	*Audio (channels, sample rate)*
E-mail	160x120	9.99	Mono, 22.05 kHz
Web	240x180	11.99	Stereo, 22.05 kHz
Web streaming	240x180	11.99	Stereo, 22.05 kHz
CD-ROM	320x240	14.99	Stereo, 44.1 kHz
Full quality DV (NTSC)	720x480	29.97	Stereo, 32 kHz

Exporting Internet movies from Studio

Pinnacle Studio provides you with a number of different export options. Two Web-friendly formats available in Studio are RealVideo and Windows Media. To begin exporting your finished movie using either of these formats, choose View⇨Make Movie. Click the Stream button on the left side of the Make Movie window that appears. As you can see in Figure 14-8, the export controls will now include radio buttons for Windows Media or RealVideo. After you've chosen a format, click the Settings button. The settings available for each version are described in the next two sections.

When you're done adjusting settings, simply click the Create Web File button at the bottom of the export control. Choose a location for your movie and name it in the Save As dialog box that appears.

Using the RealVideo format

RealVideo is a popular format for online videos, and making RealVideo movies in Pinnacle Studio is pretty easy. In the Make Movie window, click Stream, and then select the RealVideo radio button. Next, click Settings. The Make RealVideo tab of the Pinnacle Studio Setup Options dialog box appears, as shown in Figure 14-9.

Review the following settings, adjusting them as needed:

- **Title:** Enter a title for your movie. This title will appear in the program window when people view your movie, so it should be written in plain English.
- **Author:** That's you! Enter your name here.

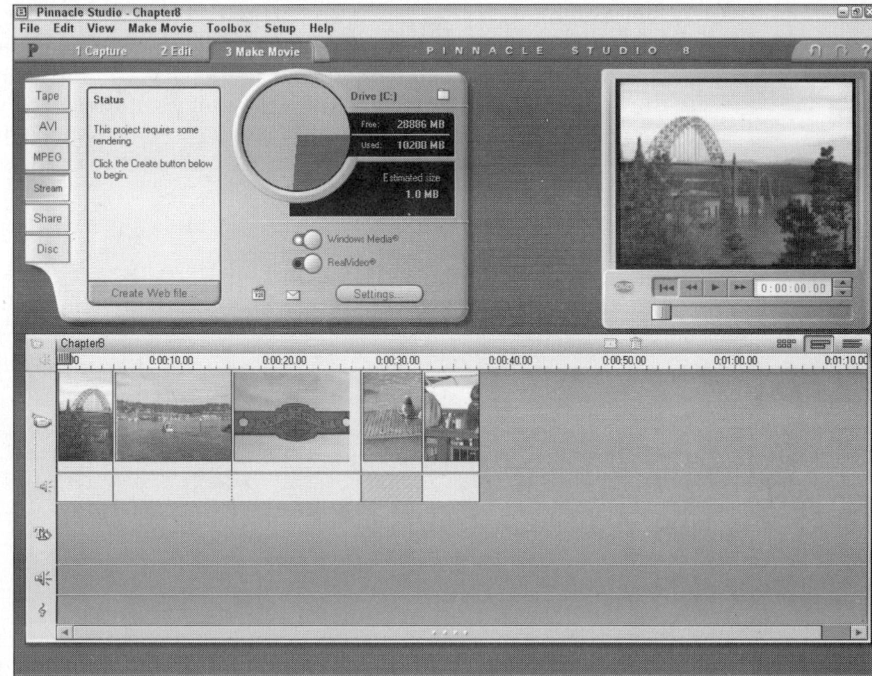

Figure 14-8: Click Stream, and then choose Windows Media or RealVideo in the Make Movie window.

- **Copyright:** Enter the year. You can enter the month and day if you wish, but it's not necessary.
- **Keywords:** Enter some keywords relating to your movie. This will help people who are searching for your movie using the keywords you list.
- **Video Quality:** Choose a video quality option here. Most of the time, the safest choice in this menu is Normal Motion Video. The Smoothest Motion Video option works well with video that doesn't have a lot of action, whereas the Sharpest Image Video option is best for video that *does* have a lot of action. The Slide Show option shows a series of still images, which obviously isn't ideal for most video. If you choose the No Video option, no video will be included in the file.
- **Audio Quality:** Choose an option from this menu that matches the majority of audio in your project. Choices include No Audio, Voice Only, Voice with Background Music, Music, and Stereo Music.
- **Video Size:** Select a frame size for your video image here.
- **Web Server:** If you know that your movie will be placed on a RealServer streaming media server, choose the RealServer option. Otherwise, choose HTTP.

- **Target Audience:** If your movie will be placed on a RealServer, you can choose multiple options here. The server will automatically detect the connection speed of each person who accesses your movie, and a movie of the appropriate quality level will be sent. If you are placing your movie on a regular Web server (HTTP), you can only choose one option. Movie quality settings will be automatically tailored to the Target audience that you choose.

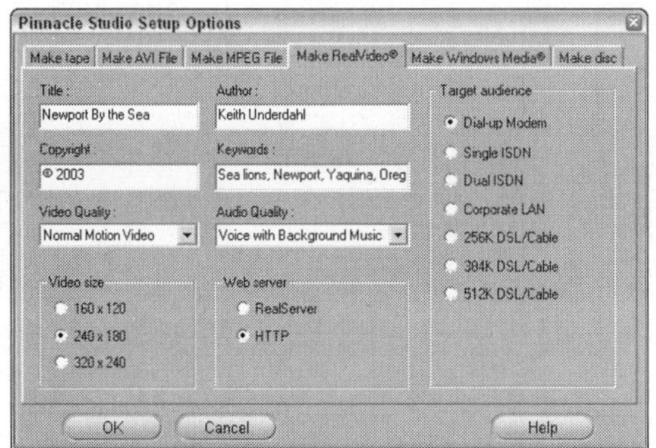

Figure 14-9: Adjust settings for RealVideo export.

If you're placing your movie on a regular Web server (HTTP), you may want to output two different versions of the same movie for people with different connection speeds. Output a lower-quality movie using the Dial-up Modem setting, and then output a higher-quality version of the same movie using one of the DSL/Cable settings. Give each file a unique name and provide separate descriptive links to each one on your Web page.

Using the Windows Media format

Although Microsoft was late to the online multimedia game, Windows Media is quickly becoming one of the most popular video formats on the Web. Pinnacle Studio can export directly to Windows Media format. To do so, choose View ➪ Make Movie, and then choose Stream on the left side of the Make Movie window. Click the Windows Media radio button that appears. I strongly recommend that you review export settings before making a Windows Media file, and you can review those settings by clicking the (surprise!) Settings button in the Make Movie window. The Make Windows Media tab of the Pinnacle Studio Setup Options dialog box appears, as shown in Figure 14-10.

When you adjust settings for Windows Media export, check the following:

- **Title:** Enter a plain English title for your movie here. This title will appear at the bottom of the Windows Media Player window when your movie is played.

- **Author:** They can't give out awards if they don't know who made the movie! Enter your name here to give yourself proper credit.
- **Copyright:** Enter a year, and month and day if you like.
- **Description:** Type a brief description of your movie. This description will scroll across the bottom of the Windows Media Player window as the movie plays.
- **Rating:** Give your movie a rating if you want.
- **Markers:** If you include Markers in your Windows Media movie, viewers can jump from clip to clip by pressing the Next and Previous buttons in their Windows Media Player programs. As you can see in Figure 14-10, I have chosen to add a marker for every clip. If you choose the Markers for Named Clips Only option, only clips that you manually named while you were editing your project (see Chapter 8) will have markers.
- **Playback Quality:** Choose the Low, Medium, or High presets in the menu on the left. If you choose Custom, a second menu appears to the right, displaying a wider selection of presets. A summary of movie settings for each preset is shown under the Playback Quality menus.

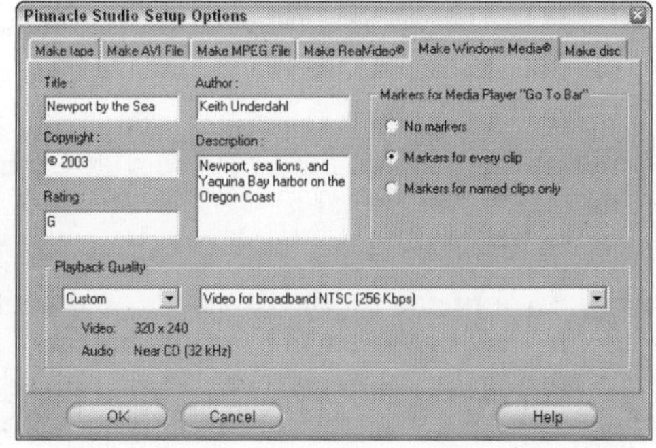

Figure 14-10: Adjust settings for Windows Media export.

Putting Your Movie on the Web

After your movies are exported in a Web-friendly format, it's time to actually make those movies available *on* the Web. This means you'll have to upload your movies to a *Web server* (basically a big hard disk that is connected to the Internet, set up so anyone with a Web browser can access files kept on it). The next two sections help you find a Web server on which to store your movie files, and I show you how to make a simple Web page that can serve as your online theater.

Finding an online home for your movies

If you want other people to be able to download and watch your movies, you'll need to put the movie files on a Web server. Your Internet service provider (ISP) might actually provide some free Web server space with your Internet account. This free space is usually limited to about 5 to 15MB, but the exact amount varies greatly. You can use your Web server space to publish pictures, movies, and Web pages that anyone on the Internet can see. Check with your ISP to find out whether you have some available Web server space — and, if you do, get instructions for uploading your files to their Web server.

If your ISP doesn't provide Web server space, or if it isn't enough space to hold all of your movie files, don't worry. Plenty of other resources are available. Several companies specialize in selling server space that you can use to store your movies. Three services include

- **.Mac (`www.mac.com`):** This service from Apple includes e-mail tools, an address book, antivirus service, and most importantly, 100 MB of storage space on their Web server. Uploading movie files to .Mac is just as easy as copying files to different disks on your computer. The .Mac service costs approximately $100 per year, and provides many more features than I can list here.
- **HugeHost.com (`www.hugehost.com`):** As the name implies, HugeHost.com lets you put huge files online. The service is quite affordable, as well. For example, 1000MB (yes, one *thousand*) is just $5 per month, or $55 per year. See their Web site for other pricing plans.
- **Neptune Mediashare (`www.neptune.com`):** This service is partnered with Microsoft so you can easily access the Neptune Web site directly from within Windows Movie Maker. When you export a movie for the Web from Movie Maker, you are given the opportunity to log on to your Neptune Mediashare account and upload files instantly. The Mediashare Pro service provides 100MB of storage space for $39 per year.

Whatever you use as a Web server for your movie files, make sure you get specific instructions for uploading. You'll also need to know what the Web address is for the files that you upload. You can then send that address to other people so that they can find and download your movie.

Creating a (very) simple Web page

Yes, you read that heading correctly, I'm going to show you how to create a simple little Web page. Yes, this is still *Digital Video For Dummies*, and no, I'm not going to tell you *everything* you need to know about designing and managing a Web site. But making a simple little Web page is actually pretty, um, simple. To make a Web page, open a text-editing program:

- **Macintosh:** Open the Apple Tools folder in your Applications folder, and then double-click the TextEdit icon.
- **Windows:** Choose Start⇨All Programs ⇨ Accessories ⇨ Notepad.

After you have your text-editing program open, type the following exactly as shown here:

```
<html>
<head>
<title>My Online Theater</title>
</head>
<body>
<center>
<h1>My Online Theater</h1>
<p><a href="Newport.wmv">Newport By the Sea</a> 1.03 MB,
           Windows Media Video</p>
</center>
</body>
</html>
```

You'll notice that the line which starts with `<p>` refers to the name of a movie that I created. The text between the quotation marks (`Newpowt.wmv`, in my example) is the filename of the movie file. Change that text to match the filename of your own movie. Filenames on Web sites are usually case sensitive, so make sure you use upper- and lowercase letters exactly as they are used by the movie file. Also change the descriptive text (`Newport By the Sea`, in my example) part to match your own movie. I've also listed the file size and format, just so people who visit the page will know what to expect. You can change that information on your own file as well. When you're done typing all these lines, save the file and give it the following filename:

```
index.html
```

Upload this file to the same directory on your Web server that contains your movie file. (You should get upload instructions from your ISP or Web server provider.) Make sure that both this `index.html` file and your movie file are in the exact same directory on the Web server. If the address to your movie file is:

```
www.webserver.com/myaccount/Newport.wmv
```

Simply provide your audience with this address instead:

```
www.webserver.com/myaccount/
```

When people visit that page, they'll see something similar to Figure 14-11. If you have several movies on your Web server, you can list them all on a single page. Just copy that line that starts with `<p>`, and then enter the filename and other information for each movie in each respective line.

There's a lot more to Web page creation and management than what I have described here. Web design is a big subject, and if you're interested in creating more complex Web pages, I recommend that you obtain a book that covers the subject more thoroughly, such as *HTML 4 For Dummies,* Fourth Edition, by Ed Tittel and Natanya Pitts, or *Creating Web Pages For Dummies* by Bud Smith and Arthur Bebak, both published by Wiley Publishing, Inc.

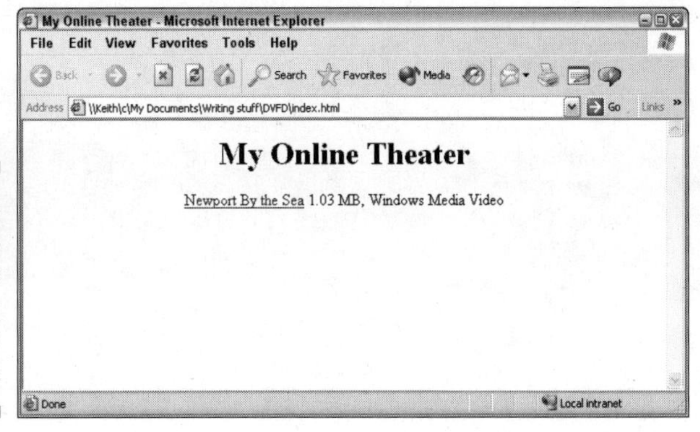

Figure 14-11: A simple Web page makes your movies easy to access.

Chapter 15

Exporting Digital Video to Tape

In This Chapter

- Preparing your movie for TV playback
- Setting up your hardware for export
- Exporting your movie

I don't think I'm going out on a limb by suggesting that almost everyone you know owns a TV and a VCR. These ubiquitous bits of home entertainment gear are all anyone needs to view your latest movie production — because exporting your movie to videotape is pretty simple. Well, it's *usually* simple, especially *if* you have the right hardware. But even if all you have is a digital camcorder, sending your movie to videotape isn't very difficult.

This chapter shows you how to get your movie ready for videotape, and helps you prepare the movie project and your export hardware. Then I show you how to actually export the movie to tape.

Prepping Your Movie for TV Playback

Throughout this book, I harp on the fact that televisions and computer monitors are very different. This means that the video that looks just peachy in the preview window of your editing software may not look all that great when it's viewed on a regular TV. Computer monitors and TVs differ in three important ways:

- **Color:** Computer monitors and television screens generate colors differently. This means that colors that look fine on your computer may not look so hot when viewed on a TV.
- **Pixel shape:** Video images are made up of a grid of tiny little blocks called *pixels*. Pixels on computer monitors are square, but the pixels in TV images are slightly rectangular. Pixel shape is discussed in greater detail in Chapter 3, but basically this means that some images that look okay on your computer may appear slightly stretched or squeezed on a

TV. This usually isn't a problem for video captured from your camcorder, but still images and graphics generated on your computer could be a problem. (See Chapter 12 for more on preparing still graphics for your movie.)

- **Interlacing:** As I describe in Chapter 3, TV video images are usually interlaced, whereas computer monitors draw images by using progressive scanning. The main problems you encounter when you export a project to tape is that the very thin lines that show up on the screen may flicker or appear to crawl. Pay special attention to titles, where thin lines are likely to appear in some letters. (See Chapter 9 for tips on avoiding this problem.)

In view (so to speak) of these issues, I strongly recommend that you try to preview your movie on an external monitor first. Check out Chapter 13 for instructions.

If you use the LCD display on your camcorder to preview your movie, keep in mind that the LCD panel probably isn't interlaced. However, the camcorder's viewfinder probably *is* interlaced. This means that flickering thin lines (for example) may show up in the viewfinder but not on the LCD panel. Preview the movie using *both* the LCD display and the viewfinder before you actually export it.

Setting Up Your Hardware

Getting your hardware ready for exporting a movie to tape isn't so difficult, really. The easiest thing to do is connect your digital camcorder to your FireWire port and turn on your camcorder to VTR or Player mode. (Oh yeah, and insert a blank tape into the camcorder.) After your movie is recorded onto the tape in your camcorder, you can connect the camcorder to a regular VCR and dub your movie onto a regular VHS tape if you want.

I strongly urge you to use a fresh tape that has black video recorded on its entire length. This will prevent errors in communication between your digital camcorder and your computer. See the sidebar "Blacking and Coding Your Tapes" in Chapter 4 for instructions on how to prepare a digital videotape for use.

If your master plan is to eventually record your movie on a VHS tape, you may want to skip the middleman — that would be your digital camcorder — and record straight from your computer to a regular VCR. To do so, you have three basic options:

- **Use an analog video-capture card.** Analog capture cards (such as the Pinnacle AV/DV board) can usually export to an analog source as well as import from one. When you export video using an analog card, I strongly recommend you use the software that came with that card. Most analog capture cards come with special utilities to help you import and export video. The Pinnacle AV/DV board uses Pinnacle Studio to capture and export video. To get Studio ready for analog export, follow these steps:

  1. **Connect the analog outputs for the card to the video inputs on your VCR.**
  2. **Make sure the software that came with the capture card is set to export to the correct ports.**

     The Pinnacle AV/DV, for example, uses the Pinnacle Studio software. In Studio, choose Setup ⇨ Make Tape. The Pinnacle Studio Setup Options dialog box appears, as shown in Figure 15-1. On the Make Tape tab, choose Studio AV/DV analog in the Video drop-down list.

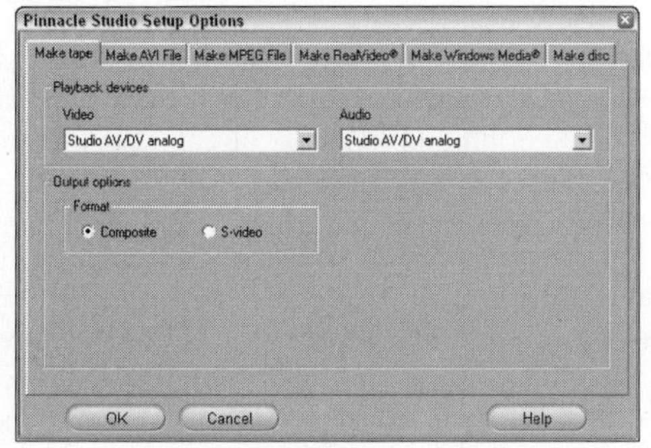

Figure 15-1: Choose analog outputs using Studio's Make Tape setup options.

  3. **Make sure that the right analog output ports are selected.**

     The Pinnacle AV/DV board has both composite and S-Video outputs, so choose the one to which you have connected your VCR.

  Composite and S-Video connectors are described in Chapter 6.

- **Use a video converter.** Chapter 6 describes how to capture analog video using a converter that connects to your computer's FireWire port. The converters listed in Chapter 6 also have analog outputs to which you can connect a VCR.
- **Use your digital camcorder as a converter.** I know, I know, I said I was going to show you how to *avoid* using your camcorder as the middleman when you export to VHS tape. But if you don't have an analog-capture

card or a video converter, you might be able to connect your digital camcorder to your FireWire port, and then connect a VCR to the camcorder's analog outputs. If nothing else, this arrangement reduces wear and tear on your camcorder's expensive tape-drive mechanism. Some digital camcorders won't allow you to make this connection, because some models can't send video out the analog ports at the same time they're taking video in through the FireWire cable. Experiment with your own camcorder and VCR and see whether this arrangement will work for you.

If you are exporting to a VCR, make sure that a new, blank tape is inserted and ready to use, and make sure the VCR is set to the right channel. (Many VCRs have to be set to a special "AV" channel to accept video from composite video cables.) As a last step before you begin your export, preview your movie on a TV connected to the VCR (see Chapter 13) to make sure that the VCR is picking up the signal.

Exporting the Movie

After your hardware is set up properly and you're sure that your movie will look good on a regular TV, you're ready to export the movie. Regardless of what software you are using, keep in mind that — like video capture — video export uses a lot of memory and computer resources. To make sure that your system is ready for export:

- **Turn off unnecessary programs.** If you're like me, you probably feel like you can't live without your e-mail program, Internet messaging program, Web browser, and music jukebox all running at once. Maybe *you* can't live without these things, but your video-editing software will get along just fine without them. In fact, the export process will work much better if these things are closed, and you're less likely to have dropped frames or other quality problems during export.
- **Disable power-management settings.** If you're exporting a movie that's 30 minutes long, and your hard disk is set to go into power-saving mode after 15 minutes, you could have a problem during export because the computer will mistakenly decide that exporting a movie is the same thing as inactivity. Power management is usually a good thing, but if your hard disk or other system components go into sleep mode during export, the video export will fail. Pay special attention to this if you're working on a laptop, which probably has pretty aggressive power-management settings right now.
    - On a Mac, use the Energy Saver icon in System Preferences to adjust power settings. Crank all the sliders in the Energy Saver window up to Never before you export your movie.

- In Windows, open the Control Panel, click the Performance and Maintenance category if you see it, and then open the Power Options icon. Set all of the pull-down menus to Never before exporting your movie.

- **Disable screen savers.** Screen savers aren't quite as likely to ruin a movie export as power-management settings, but it's better to be safe than sorry.
  - In Windows, right-click a blank area of the Windows desktop and choose Properties. Click the Screen Saver tab of the Display Properties dialog box, and choose the None screen saver. (That's my favorite one, personally.)
  - On your Mac, open the Screen Saver icon in System Preferences and choose Never on the Activation tab of the Screen Saver dialog box.

Whenever you export a movie to tape, I always recommend that you place some black video at the beginning and end of the movie. Black video at the beginning of the tape gives your audience some time to sit down and relax between the time they push Play and the movie actually starts. Black video at the end of the movie also gives your viewers some time to press Stop before that loud, bright static comes on and puts out someone's eye. Some editing programs — like Apple iMovie — have tools that allow you to automatically insert black video during the export process. I'll show you how to insert black video using iMovie in the next section. But if you're using some other software that doesn't have this feature — like Pinnacle Studio or Windows Movie Maker — you'll need to add a clip of black video to the beginning and end of the project's timeline. You can usually do this by creating a blank full-screen title, and I'll show you how later in this chapter.

The most common failures encountered on VHS tapes are mangling or breakage at the very beginning of the tapes. Employees at video rental stores are quite skilled in the art of VHS tape repair, and they often repair beginning-of-tape problems by cutting off the damaged tape, and then re-attaching the remaining good tape to the reel. If you put 30 seconds of black video at the beginning of your VHS tapes, about three feet of tape can be cut off before any of your movie is trimmed away. And if you ever need such a repair performed, head down to your local video store. You should be able to find someone there who will do the job for a couple of dollars.

Exporting to tape in Apple iMovie

Like most software from Apple, iMovie is entirely functional and to-the-point. And at no time is this more evident than when you want to export your movie to tape. iMovie doesn't have the ability to export directly to an analog capture card, but it definitely can export video at full quality to your digital camcorder. To export your finished movie to tape, follow these steps:

1. **Connect your digital camcorder to the FireWire port on your computer, and turn the camera on to VTR or Player mode.**

   Make sure that you have a new, blank videotape cued up and ready in the camcorder.

2. **In iMovie, choose File⇨Export.**

   The iMovie Export dialog box appears as shown in Figure 15-2.

3. **Choose To Camera from the Export menu.**

4. **Adjust the Wait field if you want.**

   The Wait field controls how long iMovie waits for the camera to get ready before it begins export. I recommend leaving the Wait field set at five seconds unless you're exporting to a video converter (such as the Dazzle Hollywood DV Bridge) connected to your FireWire port. In that case, you may want to increase the wait to about ten seconds or so to ensure that you have enough time to press the Record button manually on your VCR.

   Whatever you do, *don't* reduce the Wait field to less than five seconds. Virtually all camcorders need some time to bring their tape-drive mechanisms up to the proper speed, and the Wait gives the camcorder time to get ready.

5. **Adjust the two Add fields to determine the amount of black video that will be recorded at the beginning and end of the tape.**

   I recommend putting at least 30 seconds of black video at the beginning and end of the movie.

6. **Click Export.**

   iMovie will automatically export your movie to the tape in your camcorder. If you're exporting directly to a digital camcorder, iMovie will automatically control the camera for you; there's no need to press the Record button on the camcorder. But if you are exporting through a video converter, you'll need to manually press Record on your analog VCR.

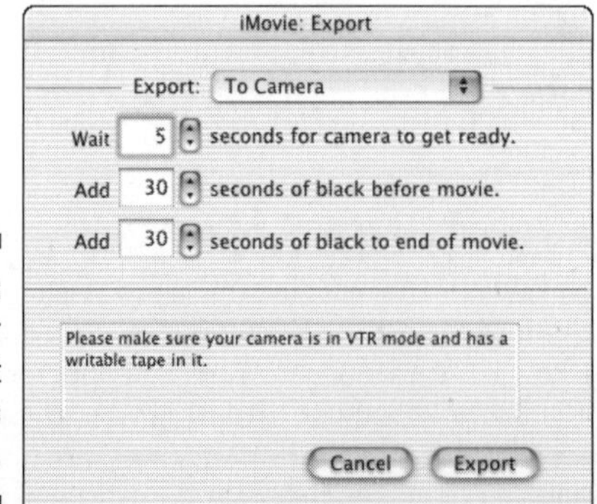

Figure 15-2: Set your export options here.

Exporting to tape in Windows Movie Maker

You can export to tape from Windows Movie Maker, but I really don't recommend it if you have another option. Even if you use Movie Maker's very highest-quality capture option, your video is still compressed by Windows Movie Maker when it's first captured into your computer. This means that the quality of the source footage stored on your computer is somewhat reduced right from the start.

If you *don't* have any other choice but to use Windows Movie Maker, here's how to do it: First connect your digital camcorder to the computer's FireWire port. I also strongly recommend that you use Movie Maker's titling tool to create a blank title with no words and a black background. Place this title at the beginning of your project and make it 30 seconds long. Copy the title and place it at the end of the project as well.

When you're ready to send your movie to tape, click Send to DV Camera under the Finish Movie step in Movie Maker (remember, I recommend using Windows Movie Maker version 2 or later; see Appendix E for information on downloading the latest version). Put a blank tape in your camcorder and follow the instructions on-screen to export your movie to tape.

Exporting to tape in Pinnacle Studio

Exporting movies to tape from Pinnacle Studio is a slightly more complex process than what is found in, say, Apple iMovie or Windows Movie Maker. But one of the main reasons for this complexity is that Studio gives you more export options than iMovie — and it exports far better-quality video than Movie Maker. The next two sections show you how to export your movie to tape from Pinnacle Studio.

Adding black video to your timeline

Earlier in this chapter, I described the importance of placing black video at the beginning and end of a movie that will be recorded onto tape. If you plan to export your Pinnacle Studio movie project to tape in a digital camcorder, you'll need to add some black video clips to the beginning and end of the timeline. To add a black video clip to the beginning of your project, follow these steps:

1. **In the Edit mode, click the Titles tab on the left side of the album.**

   A selection of titles appears in the album.

2. **If any tracks in the timeline are currently locked, click the track headers on the left side of the timeline to unlock them.**

When a track is locked, a tiny lock icon appears on the track header and a zebra-stripe pattern appears across the track. Unlocking all tracks is an important step because you're going to insert a title clip at the very beginning of the timeline. If all tracks are unlocked, they all shift over automatically when you insert the title. This keeps all your narration, music, and title overlays properly synchronized with your video.

3. **Click-and-drag any title to the very beginning of the video track on your timeline.**
4. **Double-click the title to open the title editor as shown in Figure 15-3.**
5. **Select the text in the title and press Delete on your keyboard to delete all the title text.**
6. **Adjust the duration of the title using the Duration field in the upper-right corner of the title editor.**

   I recommend a duration of 30 seconds.
7. **Close the title editor.**

   The blank title will appear at the beginning of the timeline.

Adjust duration.

Delete this text.

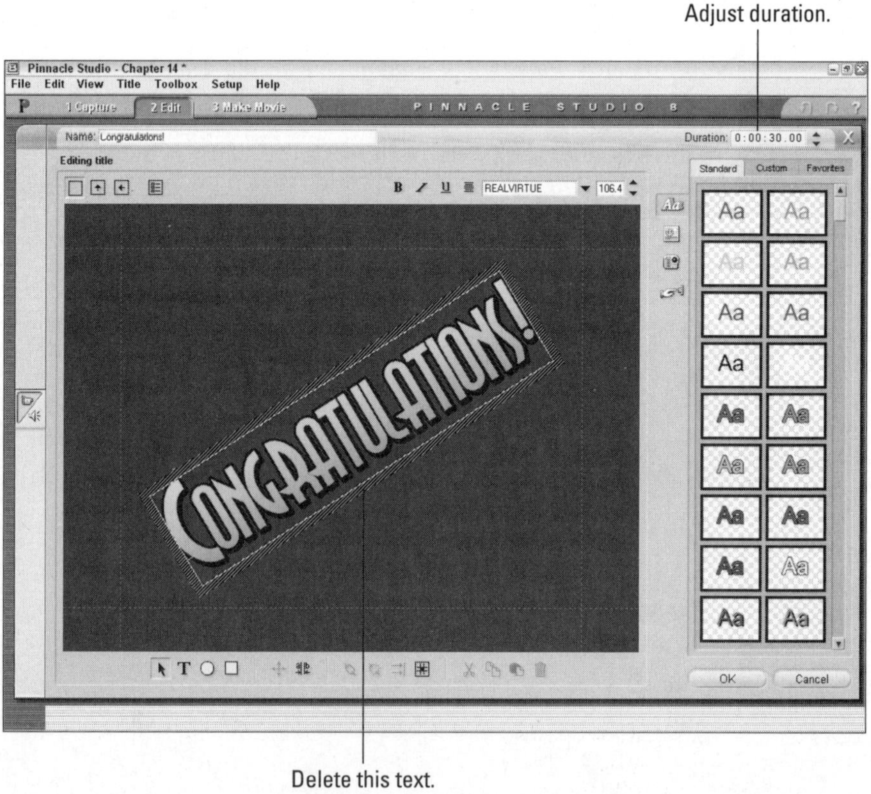

Figure 15-3: Use the title editor to create a clip of black video.

8. **Click the blank title once to select it, and then choose Edit⇨Copy.**
9. **Move the play head to the end of the timeline.**
10. **Choose Edit⇨Paste.**

    A copy of the blank title will now appear at the end of the movie as well.

Another thing I often like to do is add a Dissolve transition between the initial black video clip and the first actual clip of the movie. This technique makes the beginning of the movie a little easier on the eyes as it fades in. In Figure 15-4, I have done this by dragging a Dissolve transition to the timeline from the Transitions tab of the Studio album. See Chapter 9 for more on using transitions.

Exporting the movie

Pinnacle Studio provides the Make Movie mode as your central location for exporting a finished movie project, whether you're exporting to tape, DVD, the Internet, or carrier pigeon. (Just kidding: Export to Carrier Pigeon won't be available until the *next* version of Studio, if not later.) To open the Make Movie mode, choose View ⇨ Make Movie.

Figure 15-4: I like to add a transition between the black video and the first clip of the movie.

If any of the clips in your movie were captured at Preview quality (see Chapter 5 for details on Preview quality capture), you'll be prompted to insert the tapes containing the original source clips at full quality. Make sure you have those tapes handy, and follow the instructions on-screen to recapture the footage.

You're ready to start exporting your movie to tape. Follow these steps:

1. **In the Make Movie mode, click Tape at the left side of the Make Movie window.**

   Basic video settings will appear in the Make Movie window, as well as the estimated file size for the exported file. Studio needs to export the movie as a file before it can be recorded onto tape, and that file is probably going to be big. As you can see in Figure 15-5, I have a movie that is only about 50 seconds long and yet it will create a file that is over 157MB (megabytes). This is why I always recommend you buy the very biggest hard disk you can afford.

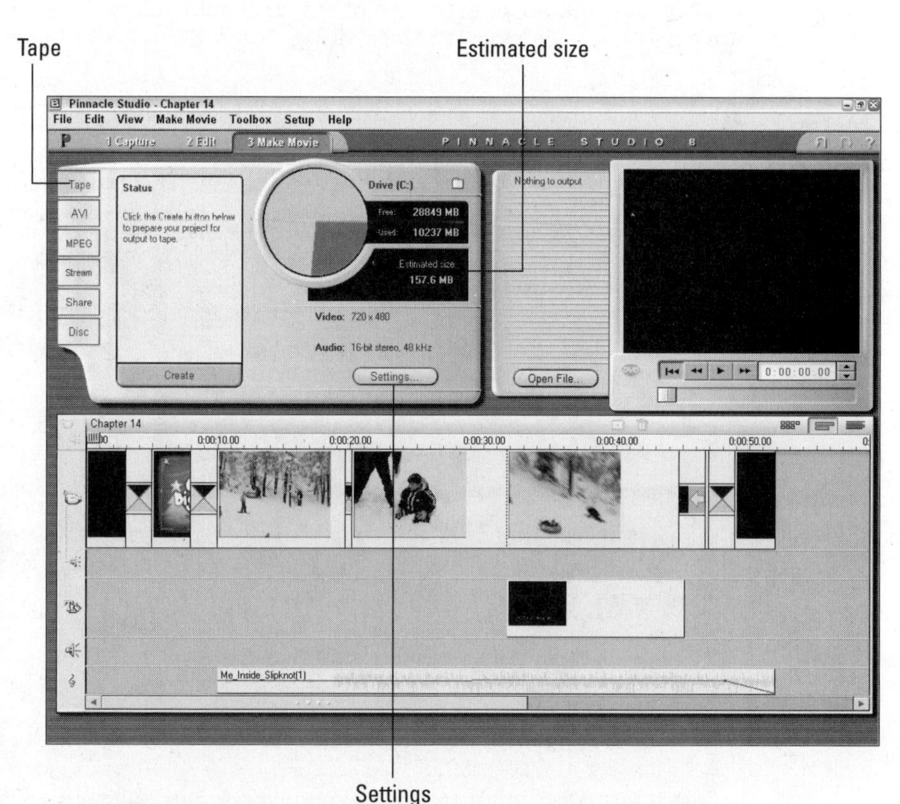

Figure 15-5: Studio is ready to export my movie to tape.

2. **Click the Settings button.**

   The Make Tape tab of the Pinnacle Studio Setup Options dialog appears. As described earlier in this chapter, make sure that the correct output source is selected in the Video menu. In Figure 15-6, you can see that I am preparing to export to a DV camcorder connected to my FireWire (IEEE-1394) port.

Figure 15-6: Review your Make Tape options here.

3. **If you're exporting to a DV camcorder, place a check mark next to the Automatically Start and Stop Recording option.**

   With this option enabled, Studio will automatically control your camcorder for you, meaning you won't have to manually press Record on the camcorder. If you're exporting to a video converter such as the Dazzle Hollywood DV Bridge, do not enable this option.

   If you do enable automatic control of your DV camera, I recommend that you set the Record Delay Time to five seconds as shown in Figure 15-6. This gives the camcorder's tape mechanism enough time to spool up to the proper speed for recording.

4. **Click OK to close the Pinnacle Studio Setup Options dialog box.**

5. **Back in the Make Movie window, click Create at the bottom of the export control.**

   Studio creates a file for your movie. The process may take several minutes, especially if your movie is long and has a lot of effects and transitions. When the file is created, the export control tells you that your project is ready for output, as shown in Figure 15-7.

6. **Click Play under the preview window.**

   If you chose to give Studio automatic control of your DV camcorder (Step 3), Studio automatically starts the recording feature on your camcorder, stopping when the movie is completely exported.

If you're exporting through an analog output (such as a Pinnacle AV/DV card), you have to press Record on your analog VCR a few seconds before you click Play in the Studio preview window. One nice thing about Studio is that while it's waiting for you to click Play, the software sends out a black video signal through the analog outputs. This means that you can press Record on your analog VCR and let it record that black video for 30 seconds or so before you play your movie in the export process. Presto — you eliminate the need to add black video clips to the beginning and end of the timeline. When the movie is done being exported, Studio reverts to outputting black video through the analog outputs. I suggest you record 20 or 30 seconds of this black video on the VHS tape before you press Stop.

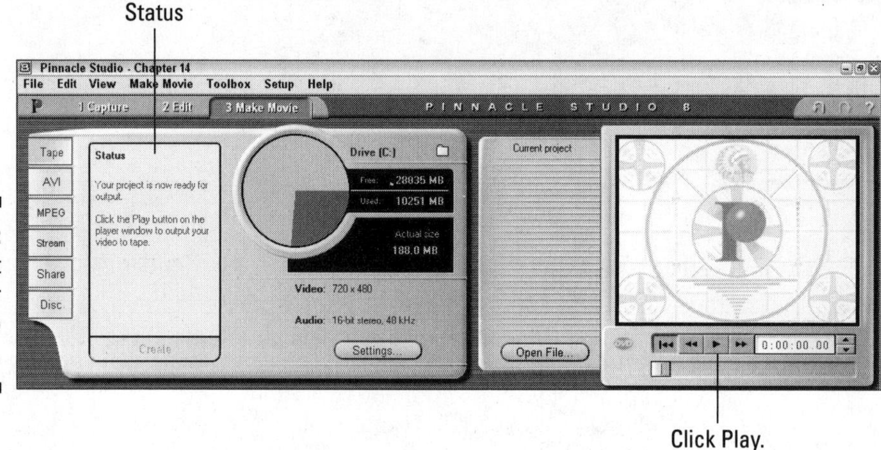

Figure 15-7: The project is ready for output to tape.

Chapter 16

Recording CDs and DVDs

In This Chapter

- Understanding DVD basics
- Making effective disc menus
- Understanding the DVD alphabet soup
- Burning DVDs

You probably don't need me to tell you that good old-fashioned videotapes are going the way of the dinosaur. They're being replaced by shiny little round platters called DVDs. A DVD (which some say stands for *digital versatile disc*, others think it's *digital video disc*, and which others contend that it stands for nothing at all) is just 5.25" in diameter (just like a music CD or CD-ROM), but can hold hours of high quality video. And as long as a DVD disc doesn't get scratched up, it will continue to provide the same quality no matter how many times it is played.

Until recently, the process of recording DVDs was complex and very expensive. In October 2000 I managed a project to master a software DVD. The recording equipment cost thousands of dollars, and we paid $40 each for blank recordable DVD discs. But just three years later DVD recorders can be had for less than $300, and blank discs cost about $1 each, maybe less if you shop around.

This chapter shows you how to put your movies onto DVD. And if you don't have a DVD burner yet, I'll show you how to make discs in a regular CD burner that will play short movies in most DVD players.

Understanding DVD Basics

Movies on DVD have been around since the mid-1990s, and DVDs are quickly replacing VHS tapes as the de facto standard for video distribution. DVDs provide higher quality and are physically smaller than VHS, and unlike tapes, they don't have moving parts that break after a number of uses. Best of all, your local video-rental store will never charge you a fee if you forget to rewind a DVD.

If DVDs have one disadvantage compared to videotapes, it's that the process of recording a movie onto DVD is a little more complicated. The following sections help you prepare to make your own DVDs by showing you what you need — and translating some terms and acronyms used in the DVD world.

Getting ready to record DVDs

If all you wanted to do was put your video on VHS tapes, you could probably get away with not using a computer (provided you didn't want to do any fancy editing). But if you want to make your own DVDs, a computer is pretty much mandatory. In addition to your computer, you also need

- **A DVD burner:** Drives that can record DVDs are now widely available for just a couple hundred dollars. You can buy an internal drive that replaces your current CD-ROM or CD-R drive, but installing internal drives requires some level of computer hardware expertise. An easier solution is to buy an external DVD burner that connects to your computer's FireWire or USB 2.0 port. The easiest solution of all (but also the most expensive) is to buy a new computer with a DVD burner already installed. Many new Macintosh computers come with the SuperDrive, which is Apple's DVD burner.
- **Blank DVDs:** Blank recordable DVDs look like blank recordable CDs, but they are different (not to mention more expensive). Make sure you buy blank DVDs that are compatible with your particular DVD burner. For example, some blank discs may be compatible with DVD-R drives, but not DVD+R drives. I'll explain the difference in the next section.
- **DVD recording software:** Also called *mastering software* or *authoring software*, you need software on your computer that can properly format your video and record it onto a DVD. If you have a DVD burner, it probably came with recording software. Apple offers its own DVD recording program called iDVD (iDVD only works with Apple's SuperDrive), and Pinnacle Studio has the built-in ability to author DVDs as well.

  DVD recording software usually includes tools to help you create DVD menus for your discs. Menus allow viewers to find and use the various features of your DVD movie. Figure 16-1 shows a menu that I made for one of my DVDs.

Comprehending DVD standards

One of the things I love about VHS tapes is that after I record a movie onto a tape, I know it'll play in just about any VCR. Likewise, I can usually look at the tape and immediately know how much video it will hold. For example, a tape labeled T-120 is going to hold about 120 minutes of video. Alas, DVDs are a little more complicated. Although many DVD players will be able to play the

DVDs that you record yourself, some players will have trouble with them. And of course, the amount of space on a blank DVD is usually listed computer-style (gigabytes) rather than human-style (minutes). If a blank DVD says it can hold 4.7GB, how many minutes of video is that, exactly? The next few sections answer the most common questions you'll have about recording DVDs.

Figure 16-1: DVD authoring software helps you create DVD menus such as this.

How much video can I cram onto a DVD?

A standard recordable DVD of the type you are likely to record yourself has a capacity of 4.7GB, which works out to a little over two hours of high-quality video. Two hours is an approximation; as I show later in this chapter, quality settings greatly affect how much video you can actually squeeze onto a disc. Some professionally manufactured DVDs can hold more because they are double-sided or have more than one layer of data on a single side. Table 16-1 lists the most common DVD capacities.

Table 16-1 DVD Capacities

Type	*Capacity*	*Approximate Video Time*
Single-sided, single-layer	4.7GB	More than 2 hours
Single-sided, double-layer	8.5GB	4 hours
Double-sided, single-layer	9.4GB	4.5 hours
Double-sided, double-layer	16GB	More than 8 hours

You've probably seen double-sided DVDs before. They're often used to put the widescreen version of a movie on one side of the disc, and the full-screen version on the other. Unfortunately, there is currently no easy way for you to make double-sided or double-layer DVDs in your home or office. These types of discs require special manufacturing processes, so (for now) you're limited to about two hours of video for each DVD you record.

Choosing external DVD burners

If you don't feel like tearing into the internals of your computer to install a DVD burner, you may want to consider an external DVD burner. External DVD recording drives typically connect to either a FireWire or USB 2.0 port. If your computer already has a FireWire port, a FireWire DVD burner is the safer choice. External FireWire DVD burners are available for both Windows PCs and Macintosh computers.

Some external DVD burners can use a USB 2.0 port instead of a FireWire port. To use a USB 2.0 DVD burner, your computer must have a USB version 2.0 port. If your computer was made in the spring of 2002 or earlier, there is a good chance that it only has USB 1.1. USB 1.1 isn't fast enough to effectively burn DVDs. Check the documentation for your computer if you aren't sure. You can add USB 2.0 ports to your computer using a USB 2.0 expansion card. But if you're going to go to that much trouble, you might as well add an internal DVD burner instead. Consult a computer hardware professional about upgrading any internal parts of your computer.

When double-layer discs are manufactured, the layers are actually recorded separately and then glued together (yes, really) using a special transparent glue. This is a very complex process, so don't try to make your own double-layer DVDs with super glue; it won't work.

What is the deal with the DVD-R/RW+R/RW alphabet soup?

When it comes to buying a drive to record DVDs, you're going to see a lot of similar yet slightly different acronyms thrown around to describe the various formats that are available. The basic terms you'll encounter are

- **DVD-R (DVD-Recordable):** Like a CD-R, you can only record onto this type of disc once.
- **DVD-RW (DVD-ReWritable):** You can record onto a DVD-RW disc, erase it later, and record something else onto it.
- **DVD-RAM (DVD-Random Access Memory):** These discs can also be recorded to and erased repeatedly. DVD-RAM discs are only compatible with DVD-RAM drives, which pretty much makes this format useless for movies because most DVD players cannot play DVD-RAM discs.

The difference between DVD-R and DVD-RW is simple enough. But as you peruse advertisements for various DVD burners, you'll notice that some drives say they record DVD-R/RW, while others record DVD+R/RW. The dash (-) and the plus (+) aren't simply a case of catalog editors using different grammar. The -R and +R formats are unique standards. If you have a DVD-R drive, you must make sure that you buy DVD-R blank discs. DVD-R drives are typically made by Apple Computer, Hitachi, Panasonic, Pioneer, NEC, Toshiba, Samsung, or Sharp.

Likewise, if you have a DVD+R drive, you must buy DVD+R blank discs. Manufacturers that offer DVD+R drives include Dell, Hewlett-Packard, Philips, Sony, and Yamaha.

The differences between the -R and +R formats are not major. Either type of disc can be played in most DVD-ROM drives and DVD players. And as of this writing, both formats are widely available. I do not recommend one over the other, although I have found that blank +R discs seem to be a little more expensive.

One more thing: When you buy DVD-R discs (that's *-R*, not +R), make sure you buy discs that are labeled for General use, and not for Authoring. Not only are the DVD-R for Authoring discs more expensive, they're not compatible with most consumer DVD-R drives. This shouldn't be a huge problem because most retailers only sell DVD-R for General discs, but it's something to double-check when you buy blank media.

What are VCDs and SVCDs?

You can still make DVD movies even if you don't have a DVD burner.

Yes, you read that correctly. Okay, technically you can't make *real* DVDs without a DVD burner, but you can make discs that have menus just like DVDs and play in most DVD players. All you need is a regular old CD burner and some blank CD-Rs to make one of two types of discs:

- **VCD (Video CD):** These can hold 60 minutes of video, but the quality is about half that of a DVD.
- **SVCD (Super VCD):** These hold only 20 minutes of video, but the quality is closer to (though still a little less than) DVD quality.

These two formats are usually options you can choose in some DVD-burning programs, including Pinnacle Studio. You create VCDs and SVCDs just as you would DVDs. (Later in this chapter, I show you how to create a VCD or SVCD instead of a regular DVD.)

The advantage of VCDs and SVCDs is that you can make them right now if you already have a CD burner. Just keep in mind that VCDs and SVCDs are not compatible with all DVD players.

A great resource for compatibility information is a Web site called VCDHelp.com (`www.vcdhelp.com`). Check out the Compatibility Lists section for compatibility information on specific brands and models of DVD players.

Making Effective Menus

If you've ever watched a movie on DVD, you're probably familiar with DVD menus. When you first put a movie disc into a DVD player, a screen usually appears and offers help with navigating the various features of the disc. This menu screen usually includes links that play the movie from the beginning, jump to a specific scene, show you the special features, or change languages and other settings. Menus are an important part of any DVD, and DVD *authoring software* includes tools to help you make your own menus. When you make a menu, you need to

- Design the general appearance of the menu.
- Create buttons for the menu.
- Link menu buttons to parts of your movie.

In the next couple of sections, I show you how to make effective menus.

Creating a menu

The exact steps for making a menu vary (depending on the program you're using), but the basic process is the same. In Pinnacle Studio, for example, you start by choosing a basic menu from the Menus portion of the album, and then edit that menu using the Studio Title/Menu editor. To use Studio to create a menu for your movie, follow these steps:

1. **When you are done editing your movie, place the play head at the very beginning of the timeline.**
2. **Click the Show Menus tab on the left side of the Album and preview the predesigned menus that appear.**

   A selection of predesigned menus appears, as shown in Figure 16-2. Click a menu to preview it in the Preview window. Some menus have an animated background, and others are static. Studio offers several pages of menus; be sure to click the arrows in the upper-right corner of the album to see all the available menus.
3. **After you have selected a basic menu, drag-and-drop it to the beginning of the video track on the timeline.**

   When you drop the menu at the beginning of the timeline, Studio asks whether you want to automatically create links to each scene that comes after the menu. If you click Yes, a button is created for every video clip on the timeline. If you want to control the buttons yourself, choose No. The background image for the menu appears on the video track, and the control buttons for the menu appear in the title track.

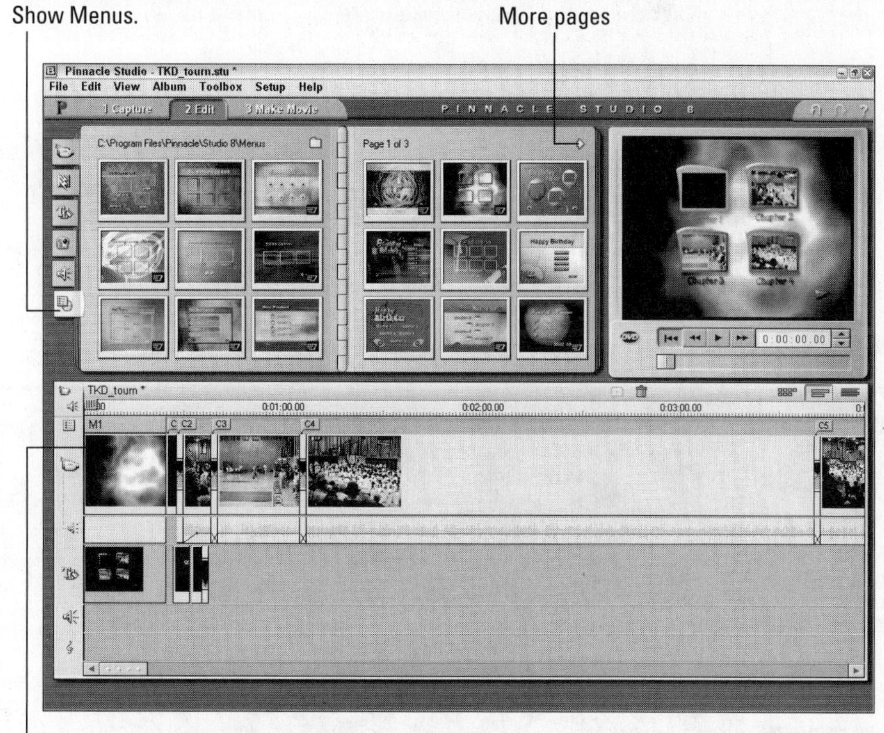

Figure 16-2: Choose a menu and drag it to the beginning of your timeline.

4. **Double-click the menu in the title track.**

   The Edit Disc Menu window appears above the timeline, as shown in Figure 16-3.

5. **If your menu has more than one page, use the Show Next Page and Show Previous Page buttons at the top of the Edit Disc Menu.**

6. **To edit a button, click it once in the Edit Disc Menu.**

   The selected button appears as a different color in the menu preview window.

7. **Enter a new name in the Name field for the button if you want.**

   In Figure 16-3, I have entered the name *Tournament* for button number 1.

8. **(Optional) If you want to control the exact point in a clip to which a menu button links, move the play head to that spot in the timeline.**

   If you chose to let Studio automatically create links to each clip, you don't need to create the links manually. If you don't want to change any current links, you can skip ahead to Step 10.

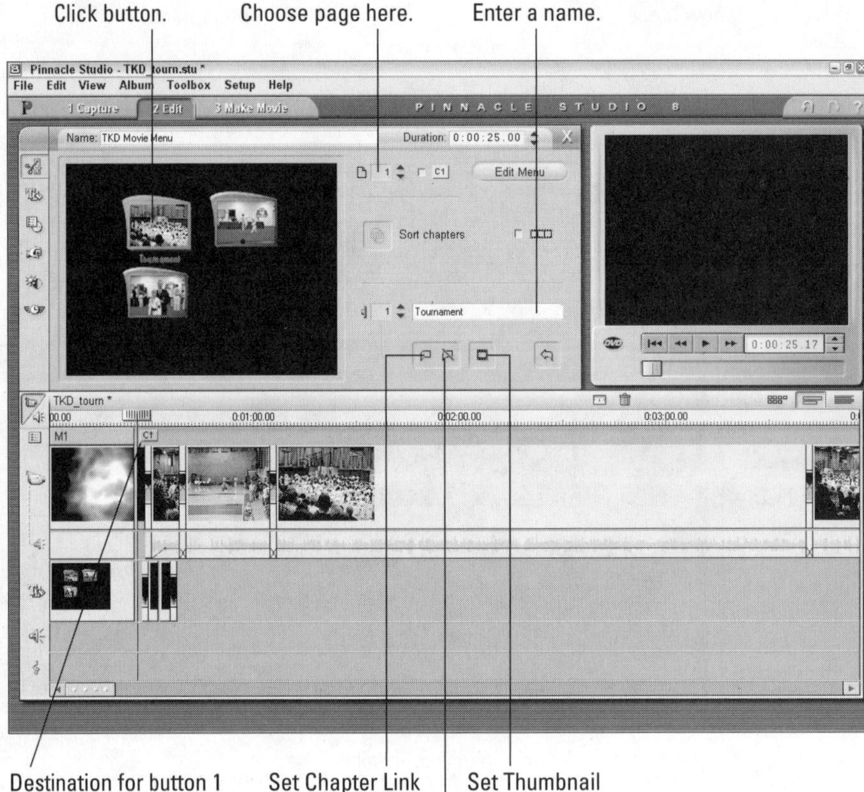

Figure 16-3: Use the Edit Disc Menu window to modify the links in your menu.

9. **Click the Set Chapter Link button.**

   A flag appears at the top of the timeline (as shown in Figure 16-3), indicating the destination for the button. Later, when users click the button you're now editing, they will be taken to this exact spot in the timeline. Click Clear Chapter link if you want to remove the current link for a button.

10. **Now move the play head to a frame that you think is representative of the button you are currently editing.**

11. **Click the Set Thumbnail button in the Edit Disc Menu window.**

    The thumbnail image for the button will now show the current frame.

12. **If you want the movie to return to the menu after a section of your movie has played, move the play head to the return spot as shown in Figure 16-4.**

13. **Click the Return to the Menu button.**

    A flag labeled M1 appears in the timeline, meaning that when the movie gets to this frame, it returns to page 1 of the menu.

    In the examples shown here, I am working on a movie project that contains several separate parts. The first part shows a Taekwondo tournament that my kids attended; the second and third parts show tests. Although all three parts will be burned onto the same DVD, my plan is for my audience to watch each one separately. So, when the tournament portion of the movie ends, I want the menu to appear again.

Figure 16-4: Set a marker to return to the menu when a certain point in the movie is reached.

14. **Repeat Steps 6 through 13 for each button in the menu.**

    In Figure 16-4, you can see that button number 2 is linked to a frame right after the end of the previous section.

15. **When you have created and linked your menu buttons, you can modify the cosmetic appearance of the menu by clicking the Edit Menu button in the Edit Disc Menu window.**

    The Studio Title/Menu editor appears, as shown in Figure 16-5.

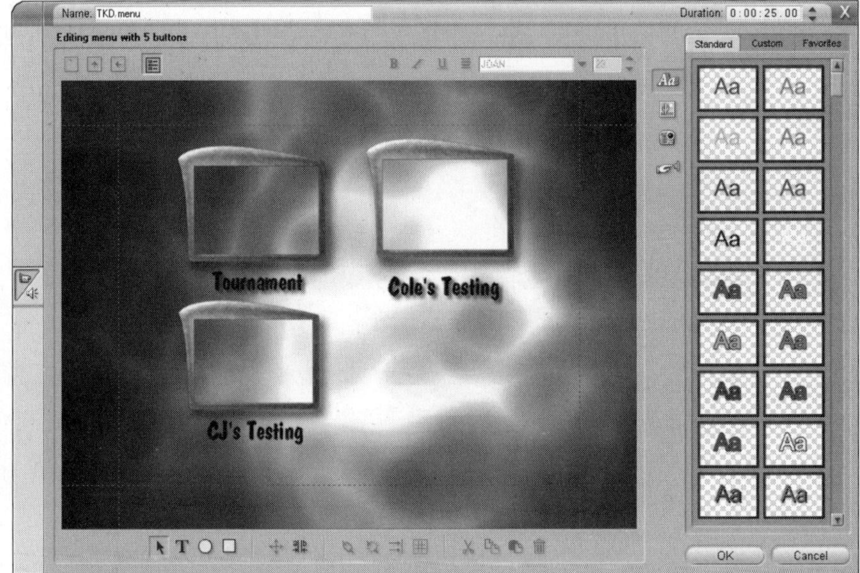

Figure 16-5: Use the Title/Menu editor to modify the appearance of your menu.

16. **Change the text and buttons as you want.**

    You edit menus in Studio the same way you edit titles. (For more on modifying objects and text properties in the Title/Menu editor, see Chapter 9.)

17. **When you're done changing the appearance of your title, click the Close (X) button in the upper-right corner of the Title/Menu editor.**

That's it! (Seventeen steps later . . .) Your menu will only appear at the beginning of the timeline, but when you record the movie onto a DVD, Studio automatically configures the disc so the menu works properly in most DVD players.

Previewing your menu

You can preview the function of your menu right in Studio. To do so, click the little DVD button in the Studio preview window. The controls at the bottom of the preview window change to resemble typical DVD controls, as shown in Figure 16-6. The controls include

- **Main menu:** Opens the first menu page for the movie.
- **Previous menu:** Opens the last menu page that was accessed.
- **Previous/Next chapter:** Jumps back and forth through the *chapters* (another name for clips) in your movie.
- **Cursor controls:** These move the on-screen cursor from one on-screen button to the next, working much like the arrow buttons on a DVD player's remote. Click the center cursor-control button to click the currently selected on-screen button.

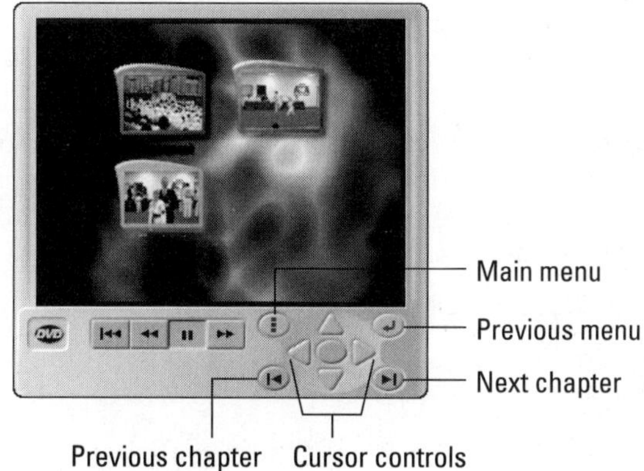

Figure 16-6: Use the Title/Menu editor to modify the appearance of your menu.

Burning DVDs

Don't worry; you won't need to keep a fire extinguisher handy for this section. *Burning* CDs or DVDs actually means recording the discs (presumably because a laser is involved in the recording process). Although "burn" is slang, it's a term commonly used by computer hardware gurus, ad-copy editors, and writers of *For Dummies* books. Affordable CD burners have been available for several years now, and affordable DVD burners are now widely available as well.

If you have a CD burner but not a DVD burner, that's okay. You can still follow the steps shown here to create a VCD or SVCD instead. So go ahead, fire up whatever kind of burner you happen to have and get ready to roast some movies!

Many different DVD authoring programs are available. As I mentioned previously in the "Making Effective Menus" section, the exact steps vary depending on the program you're using, but you should see basic settings and options similar to those shown here for Pinnacle Studio. When you're done editing your movie and have created a menu, you're ready to burn a DVD. In Studio, place a blank recordable DVD of the proper type in your DVD burner and follow these steps:

1. **Choose View⇨Make Movie.**

   The Make Movie window appears.

2. **Click Disc on the left side of the Make Movie window.**

   The create disc mode appears, as shown in Figure 16-7. In this mode, you will see two pie charts. The one in the middle indicates the amount of free space (compared to the amount of used space) on your hard disk. The pie chart on the right — the one that looks suspiciously like a DVD disc — reflects the amount of space that will be used on your disc when you burn it.

   Earlier in the chapter, I mention that a DVD-R can hold up to two hours of video. But you can fill the disc with far less video than that if you specify a higher quality setting. The amount of video that can be stuffed onto a disc is greatly affected by the quality settings you choose.

3. **To adjust settings for your disc, click the Settings button.**

   The Make Disc tab of the Pinnacle Studio Setup Options dialog box appears, as shown in Figure 16-8.

Figure 16-7: Studio's Make Movie window has all the tools you'll need for burning a DVD.

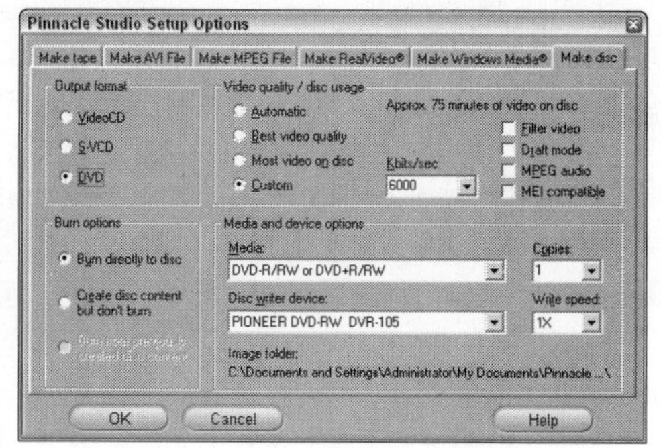

Figure 16-8: Set quality and format options here.

4. **Use the radio buttons under Output Format to choose an output format.**

   If you have a DVD burner choose DVD. Otherwise, choose VideoCD (to make a VCD) or S-VCD. Keep in mind that although VCDs and SVCDs only require a CD burner, you won't be able to put as much video on the disc, and the quality will have to be lower.

5. **Click one of the radio buttons under Video Quality/Disc Usage to choose a Video quality option.**

   My experience is that Studio's Automatic option doesn't work very well. Although the movie shown in the figures in this chapter was only 35 minutes long, Studio "automatically" chose a quality level so high that the 35-minute movie wouldn't fit on a DVD. (If your movie will not fit on the disc using the current quality settings, Studio will display a warning that the wrong type of media is in your DVD-R drive and the disc tray will try to eject your blank disc.)

   I recommend you choose Custom, and then select a number from the Kbits/sec menu. This menu controls the bit rate for the disc, which is the amount of data that will be used for every second of video. A higher bit rate gives higher quality, but takes up more space. As you adjust the Kbits/sec menu, you'll notice that an estimate of the amount of video that will fit on your disc will appear at the top of the dialog box. In Figure 16-8, I have chosen a bit rate of 6000 Kbits/sec, which will allow approximately 75 minutes of video to fit onto the disc. I recommend that you choose the highest bit rate that will allow your movie to fit onto the disc. If the bit rate is too high, when you close the Setup Options dialog box, Studio warns you that the wrong type of media is in the DVD-R drive, and the disc tray will automatically eject. If this happens, close the disc tray, click the Settings button again, and choose a lower bit rate.

6. **Choose one of the radio buttons under Burn Option.**

   If you just want to burn your movie directly onto the disc, choose Burn Directly to Disc. If you choose the second option, a copy of the disc is recorded onto your hard drive. Keep in mind that the second option will consume quite a bit of free hard disk space, possibly as much as 4.7 GB. If you create the disc content using the second option, you can later burn the disc using the third option, Burn from Previously Created Disc Content.

7. **From the Media drop-down list, choose the medium that matches the type of disc you're using.**
8. **From the Copies drop-down list, choose how many copies you want to record.**

   Think carefully, now: How many of your friends and relatives will want a copy of your movie? Don't forget to make one for yourself, too!
9. **From the Disc Writer Device drop-down list, choose the make and model of your DVD or CD burner.**

   If your burner doesn't appear in the menu, it probably isn't installed properly. If it's an external drive, make sure the power is turned on. If that doesn't work, try shutting down Studio and restarting the program. Still no luck? Check the documentation for your drive and make sure it's installed properly. You might want to try putting a regular CD-ROM or DVD into the drive to make sure it can read data.
10. **From the Write Speed drop-down list, choose how fast you want the drive to write to your disc.**

    The default speed is 1X. Even though 1X is the slowest speed, it's what I normally use. If your DVD burner supports a faster speed (say, 2X or 4X), you may be able to choose a faster speed and save some time. Just make sure that the blank disc you're using also supports the speed you choose. If you find that the discs you burn at higher speeds don't work a lot of the time, try burning at a slower speed.
11. **Click OK to close the Setup Options dialog box.**
12. **Review the message in the Status window.**

    The Status message should read, "Click the button below to start." If it says something else, such as, "Your movie is too big to fit on the disc," click the Settings button again and tweak your video-quality settings until your movie will fit.

    If your movie is more than two hours long, it's probably just plain *too long* to fit on the disc. You'll need to break the movie up into two parts and burn each part on a separate disc.
13. **Click the Create Disc button.**

    If you captured any of your video using Preview Quality Capture (see Chapter 5 for details), you'll need to connect your camcorder to your computer and insert the tape that contains the original footage. The footage will be captured at full quality.

Next, Studio starts rendering. *Rendering* is the process of compressing your movie — including all its effects, transitions, and edits — into the format used on the DVD disc. The resulting *render files* are written to your hard disk. The rendering process may take a while if your movie is long. Computers with more memory and faster processors usually render faster.

When rendering is complete, the movie is burned onto disc. When the burning process is done, a friendly message appears, telling you, "Burning Disc Completed." Your fresh-from-the-oven DVD is ejected and ready for use. Enjoy!

Part V
The Part of Tens

In this part...

I don't know about you, but my life seems to progress from one list to another — from a list of errands to run to a grocery list, to a list of scenes I want to use in my next movie. This part of *Digital Video For Dummies, 3rd Edition* provides some lists as well: ten great moviemaking tips and tricks, ten video tools you won't want to do without, and ten different video-editing programs, compared feature for feature.

Chapter 17

Ten Videography Tips and Tricks

In This Chapter

- Shooting special effects
- Shooting better video
- Dealing with environmental conditions

Throughout this book I have shown you how to shoot and edit better digital video. Every page of this book contains some nugget of great advice, but some subjects just don't seem to fit anywhere else. So, like making Friday's dinner with the week's leftovers out of the fridge, I've assembled ten leftover tips and tricks into this chapter. Read through and you're bound to find some ideas that you can use in your own movies!

Most of the tips and tricks in this chapter encourage you to use your creativity with your own video material. However, I have also included a couple of sample clips for this chapter on the CD-ROM that accompanies this book. If you want, you can use these sample clips to experiment with some of the tricks to see how they work. See Appendix A for more on accessing the sample clips on the CD-ROM.

Beam Me Up, Scotty!

Think of all the special effects you've ever seen in movies and TV shows. I'll bet that the most common effect you've seen is where a person or thing seems to magically appear or disappear from a scene. Sometimes a magician snaps his finger and blinks out of the picture. Other times people gradually fade in or out, as when crews on *Star Trek* use the transporter.

Making people appear and disappear from a scene is surprisingly easy. All you need is a camcorder, a tripod, and some simple editing software. Basically you just position the camcorder on the tripod and shoot *before* and *after* scenes. The subject should only appear in one of the scenes. The two BeamMeUp sample clips on the companion CD-ROM illustrate before and after clips.

Once you've recorded and captured your "before" and "after" clips, edit them into the timeline of your video-editing program (as I have done in Figure 17-1). If you don't use a transition between the clips, the subject will appear to "pop" into or out of the scene. If you want the subject to fade into the scene, apply a Dissolve transition between the two clips (see Chapter 9 for more on using transitions).

Figure 17-1: Make your subject magically appear (or disappear)!

To make a subject magically appear or disappear effectively, follow these basic rules:

- **Use a tripod.** A tripod is absolutely mandatory to make this effect work. You won't be able to hold the camera steady enough by hand.
- **Don't move the camera between shots.** The camera must remain absolutely still between the before and after shots. If you have to reposition the camera, or if someone bumps it, reshoot both scenes. If your camcorder has a remote control, use it to start and stop recording so that your finger doesn't move the camera at all.
- **Shoot the "before" and "after" scenes quickly.** If you're shooting outdoors and you let several minutes pass between shooting the "before" and "after" scenes, shadows and lighting might change. Even subtle light changes will be apparent when you edit the two scenes together later.

- **Don't disturb the rest of the scene.** If your subject moves a chair or picks up an object between the "before" and "after" shots, the scenes will appear inconsistent when edited together.

Seeing Stars

So you want to shoot your own science-fiction epic? All you need is a script, some willing actors, a few props from the local toy store, and you're ready to make your futuristic movie.

Well, you're *almost* ready. No sci-fi movie would be complete without a scene of spaceships flying through space, and that means you'll have to create a field of stars to serve as a backdrop. You can put stars behind your spaceships using one of two methods:

- **Edit the backdrop in later.** Most professional moviemakers shoot their spaceship models in front of a blue or green screen. The video image of the spaceship is then composited over a picture of a star-filled sky during the editing process. Professionals use this method because they can use more realistic looking starfields, and, well, just because they can.

  There's more about compositing in a later section of this chapter.

- **Create a starfield backdrop.** If you don't have the software, time, or patience to build a bluescreen studio and composite your video, just shoot the spaceships in front of a starfield backdrop that you create. The best way to create a starfield is to sew sequins onto black velvet. The velvet will absorb virtually all light that falls on it, while the sequins reflect brightly and even twinkle a little bit if the material moves (if you don't want the stars to twinkle, make sure that the velvet will not be moved by a breeze or other disturbance).

Once you have your starfield taken care of, you'll probably want to show your space ships actually moving through space. Although the natural temptation is to move the space ship past the camera, you'll find that it's very difficult to move the space ship smoothly and realistically, especially if it's suspended by fishing line or some other flexible material. I recommend you move the camera instead. Hand-carry the camera past the spaceship. This allows the camera to smoothly pan with the spaceship model while giving the appearance that the ship is moving through space.

If you have a good tripod with a fluid head, you can get an even smoother shot by placing the tripod on a wheeled stand or dolly and rolling it past the subject. Don't worry about recording the sound of wheels squeaking and scraping across the floor; you can replace that with music and sound effects later.

Forcing a Perspective

We humans perceive the world as a three dimensional space. When we look out upon the world you see color and light, and you can tell which things are close to you and which things are farther away. This is called *depth perception*. Bats use a sort of natural radar for depth perception, which is fine for winged creatures that flap in the night, but we humans perceive depth using two eyes. Our eyes focus on objects, and our brains interpret the difference between what each eye sees to provide depth perception. Without two eyes, the world would look like a flat, two-dimensional place, and activities that require depth perception (say, a game of catch) would be very difficult, if not impossible.

A video camera only has one eye — which means it has no depth perception — which is why video images appear as two-dimensional pictures. You can use this to your advantage, because you can make objects look like they're right next to each other when they're actually very far apart. Video professionals use this trick often, and call it *forced perspective*.

Consider the video image in Figure 17-2. It looks like a locomotive and train cars parked in a train yard, but looks are deceiving in this case. As you can see in Figure 17-3, the locomotive in the foreground is a scale model, and the train cars are real and about 50 yards away.

Figure 17-2: A typical industrial scene?

To make forced perspective work, you must

- **Compose the shot carefully.** The illusion of forced perspective works only if the scale looks realistic for the various items in the shot. You'll probably have to fine-tune the position of your subjects and the camera to get just the right visual effect.

Figure 17-3: Nope! The camera's eye is easily deceived.

- **Focus.** If objects are very far apart, getting both of them in focus may be difficult. To control focus, follow these steps:

  1. **Set the zoom lens at the widest setting by pressing the zoom control toward "W" on the camcorder so the lens zooms all the way out.**
  2. **Turn off auto-focus (as described in your camcorder's owner's manual).**
  3. **Set the focus to infinity.**

     Some manual focus controls have an "Infinity" setting. If your camcorder does not, manually adjust the focus so objects that are 20 or more feet away are in focus.
  4. **Position the camera five to ten feet from the closer subject in your forced-perspective shot.**

     Check carefully to make sure everything is in focus before you shoot; move the camera if necessary. With most camcorders, everything beyond a distance of about five feet will probably be in focus when you zoom out and set focus to infinity.

In Figures 17-2 and 17-3, I showed you how a small object in the foreground blends well with a large object in the background. But it can also work the other way around. In fact, model-railroad enthusiasts often use forced perspective to make their train layouts seem bigger than they really are. Mountains, trees, and buildings in the background are made smaller to provide the illusion that they are farther away.

Making Your Own Sound Effects

Believe it or not, some professional videographers will tell you that sound is actually more important than video. The reasoning goes like this: A typical viewing audience is surprisingly forgiving of minor flaws and glitches in the video picture. The viewer is able to easily "tune out" imperfections, which partially explains why cartoons are so effective. However, poor sound has an immediate and significant affect on the viewer. Poor sound gives the impression of an unprofessional, poorly produced movie.

Chapter 7 shows you how to record better-quality sound. But another key aspect of your movie's audio is the sound *effects*. I don't just mean laser blasts or crude bathroom noises, but subtle, everyday sounds — not always picked up by your camcorder — that make your movie sound much more realistic. These effects are often called Foley sounds, named for sound-effects pioneer Jack Foley. Here are some easy sound effects you can make:

- **Breaking bones:** Snap carrots or celery in half. Fruit and vegetables can be used to produce many disgusting sounds.
- **Buzzing insect:** Wrap wax paper tightly around a comb, place your lips so that they are just barely touching the paper, and hum so that the wax paper makes a buzzing sound.
- **Fire:** Crumple cellophane or wax paper to simulate the sound of a crackling fire.
- **Footsteps:** Hold two shoes and tap the heels together, followed by the toes. Experiment with different shoe types for different sounds. This may take some practice to get the timing of each footstep just right.
- **Gravel or snow:** Use cat litter to simulate the sound of walking through snow or gravel.
- **Horse hooves:** This is one of *the* classic sound effects. The clop-clop-clopping of horse hooves is often made by clapping two halves of a coconut shell together.
- **Kiss:** Pucker up and give your forearm a nice big smooch to make the sound of a kiss.
- **Punch:** Punch a raw piece of steak or a raw chicken. Of course, make sure you practice safe food-handling hygiene rules when handling raw meat: Wash your hands and all other surfaces after you are done.
- **Thunder:** Shake a large piece of sheet metal to simulate a thunderstorm.
- **Town bell:** To replicate the sound of a large bell ringing, hold the handle of a metal stew pot lid, and tap the edge with a spoon or other metal object. Experiment with various strikers, lids, or other pots and pans for just the right effect.

Filtering Your Video

Say you're making a movie showing the fun people can have when they're stuck indoors on a rainy day. Such a movie wouldn't be complete without an establishing shot to show one of the subjects looking out a window at the dismal weather. Alas, when you try to shoot this scene, all you see is a big, nasty, glaring reflection on the window.

Reflections are among the many video problems you can resolve with a lens filter on your camcorder. Filters usually attach to the front of your camera lens, and change the nature of the light passing through it. Different kinds of filters have different affects. Common filter types include

- **Polarizing filter:** This type of filter often features an adjustable ring, and can be used to reduce or control reflections on windows, water, and other surfaces.
- **UV filter:** This filter reduces UV light, and is often used to protect the lens from scratches, dust, or other damage. I never use my camcorder without at least a UV filter in place.
- **Neutral density (ND) filter:** This filter works kind of like sunglasses for your camcorder. It prevents overexposure in very bright light conditions, reducing the amount of light that passes through the lens without changing the color. If you experience washed-out color when you shoot on a sunny day, try using a ND filter.
- **Color-correction filters:** Many different kinds of color-correction filters exist. These help correct for various kinds of color imbalances in your video. Some filters can enhance colors when you're shooting outdoors on an overcast day; others reduce the color cast by certain kinds of light (such as a greenish cast that comes from many fluorescent lights).
- **Soft filter:** These filters soften details slightly in your image. This filter is often used to hide skin blemishes or wrinkles on actors who are more advanced in age.
- **Star filter:** Creates starlike patterns on extreme light sources to add a sense of magic to the video.

Many more kinds of filters are available. Check with the manufacturer of your camcorder to see whether they offer filters specially designed for your camera, and also check the documentation to see what kind of filters can work with your camcorder. Many camcorders accept standard 37mm or 58mm threaded filters. Tiffen is an excellent source for filters, and their Web site (`www.tiffen.com`) has photographic samples that show the effects of various filters on your images. To see these samples, visit `www.tiffen.com` and click the link for information on Tiffen Filters. Then locate a link for the

Tiffen Filter Brochure. This online brochure provides detailed information on the filters offered by Tiffen.

Working with Assistants

Professional moviemakers are very picky about the way they work. You can learn a lot from the pros, but the reality is that most of your video "shoots" will be pretty informal. Whether you're shooting a wedding, sporting event, birthday party, or some other special event, you'll have a variety of problems that would probably horrify the typical video pro:

- You'll have little or no control over lighting and noise.
- Your subjects won't want to rehearse, much less perform multiple takes.
- You won't have a crew of professional videographers and sound engineers.
- Worst of all, your shoot won't be catered. (The horror!)

Whenever I plan to shoot video, I always try to enlist at least one assistant. Usually it's my 10-year-old son, and although he isn't the most experienced or highly trained video professional, he is a willing trouper. His help is invaluable, but before the shoot, I always brief him on the basics — be quiet, avoid bumping the equipment, and be aware of where the lens is pointed.

Handing the reins of control to someone else

Things get especially tricky when you have to enlist an assistant to actually shoot video for you. Suppose you're about to get married, and you've asked Aunt Marie to videotape the proceedings for you. To ensure that she shoots great video, you could just buy her a copy of this book and assign her to read Chapter 4. Barring that (I know how expensive weddings can be), you can give Aunt Marie a quick list of helpful tips:

- Remember that every comment you whisper to the person next to you is recorded by the camcorder's mic.
- Try not to film the back of a subject's head.
- Try to get a shot of the cake *before* we smear it all over each other.
- Find the person who is "in charge" of the venue or location and ask where you can stand for a better shot.
- Don't be afraid to move around and shoot from different locations.
- Remember, the kiss will occur right after we say our vows, so make sure you're in position to get a good shot of that.
- Thank you! You've always been my favorite aunt.

Offer these basic tips, but don't be too picky. You can always use your video-editing software later on to clean up the movie!

Rehearse!

Everybody knows what rehearsal is, but in most cases you probably don't have a script to memorize. So what is there to rehearse? You should carefully consider and plan every aspect of your video shoot. For example, if you plan to move the camera while you shoot, practice walking the path of travel to make sure there aren't any obstacles that might block you. If you're using a tripod, practice panning to make sure you can do it effectively.

Sometimes when panning a video shot, I trip on a leg of the tripod. This causes the video image to jiggle and the camera to move. Practice panning to avoid this problem.

Of course, even if you aren't shooting a scripted production, you can still have your subjects "practice" a bit before you shoot. Things to check and rehearse beforehand include these:

- If your camcorder has an audio meter, check the sound levels before you start recording. Have your subjects speak; check the levels of their speech.
- Have subjects go through the motions of the shoot, and then coach them on how to stand or move so they show up better in the video.
- Check your camera's focus. If objects in the foreground cause your camcorder's auto-focus feature to "hunt" for the correct focus, turn off auto-focus and adjust your focus manually.
- Check if the shot is overexposed. On some higher-end camcorders, an overexposed shot shows up as a zebra-stripe pattern in the viewfinder or LCD display. Otherwise you'll have to make a careful judgment and adjust exposure as necessary. Read your camcorder's documentation to see whether it lists any special exposure settings that may help you out.

To Zoom or Not to Zoom?

If there is one single mistake that almost every camcorder owner makes, it's zoom lens abuse. On most camcorders, the zoom feature is easy and fun to use, encouraging us to use it more than is prudent. The effect of constantly zooming in and out is a disorienting video image that just looks, well, amateurish.

Some professional videographers simply refuse to use the zoom lens at all while they're shooting video. I'm not quite that extreme, but there are some general zoom lens guidelines to follow on any video shoot:

- If possible, avoid zooming in or out *while* you're recording. It's usually best to adjust zoom *before* you start recording.
- If you must zoom while recording, try to zoom *only once* during the shot. This will make the zoom look planned rather than chaotic.
- Consider actually moving the camera rather than zooming in or out.
- Prevent focus hunting (where the auto-focus feature randomly goes in and out of focus) by using manual focus. Auto-focus often hunts while you zoom, but you can easily prevent this. Before you start recording, zoom in on your subject. Get the subject in focus, and then turn off auto-focus. This should lock focus on the subject. Now, zoom out and begin recording. With focus set to manual, your subject will remain in focus as you zoom in. On most camcorders, anything farther than about ten feet away will probably be in focus if you set the camera's focal control to infinity.
- Practice using the zoom control gently. Zooming slowly and smoothly is usually preferable, but it takes a practiced hand on the control.

If you have a difficult time using the zoom control smoothly, try taping or gluing a piece of foam to the zoom-slider button on your camcorder. The foam can help dampen your inputs on the control.

Dealing with the Elements

You may at times deal with extremes of temperature or other weather conditions while shooting video. No, this section isn't about making sure the people in your movies wear jackets when it's cold (although it's always wise to bundle up). I'm more concerned about the health of your camcorder right now — and several environmental factors can affect it:

- **Condensation:** If you quickly move your camera from a very cold environment to a very warm environment (or vice versa), condensation can form on the lens. It can even form inside the camera on the inner surface of the lens, which is disastrous because you cannot easily clean it. Avoid subjecting your camcorder to rapid, extreme temperature changes.
- **Heat:** Digital tapes are still subject to the same environmental hazards as old analog tapes. Don't leave your camcorder or tapes in a roasting car when it's 105 degrees outside. Professional videographers often use a cooler for storing tapes. You shouldn't use ice packs in the cooler, but simply placing the tapes in an empty cooler helps insulate them from temperature extremes.
- **Water:** A few drops of rain can quickly destroy the sensitive electronic circuits inside your camcorder. If you believe that water may be a problem,

cover your camcorder with a plastic bag, or shoot your video at another time if possible. If you actually want to show falling raindrops in your shot, you may be able to add virtual rain later while editing. Some editing programs (such as iMovie) have effects that simulate rain — sometimes more realistically than the real thing.

- **Wind:** Even a gentle breeze blowing across the screen on your camcorder's microphone can cause a loud roaring on the audio recording. Try to shield your microphone from wind unless you know you'll be replacing the audio later during the editing process.

Another environmental hazard in many video shoots is the sun — that big, bright ball of nuclear fusion that crosses the sky every day. The sun helps plants grow, provides solar energy, and helps humans generate Vitamin D. But like all good things, the sun is best enjoyed in moderation. Too much sunlight causes skin cancer, fades the paint on your car, and overexposes the subjects in your video. Natural skin tones turn into washed out blobs, and sunlight reflecting directly on your camcorder's lens causes light flares or hazing in your video image.

You'll probably shoot video outdoors on a regular basis, so follow these tips when the sun is at its brightest:

- **Use filters.** Earlier in this chapter, I describe how lens filters can improve the quality of the video you shoot. Neutral-density and color-correction filters can reduce the overexposure caused by the sun while improving your color quality.
- **Shade your lens.** If sunlight reflects directly on your lens, it can cause streaks or bright spots called *lens flares*. Higher-end camcorders usually have black hoods that extend out in front of the lens to prevent this. If your camcorder doesn't have a hood, you can make one, using black paper or photographic tape from a photographic-supply store. (Check the video image to make sure your homemade hood doesn't show up in the picture!)
- **If possible, position your subject in a shaded area.** This will allow you to take advantage of the abundant natural light without your subject being overexposed.
- **Avoid backlit situations.** If your subject is in shade and you shoot video at such an angle that the background is very bright, you'll wind up with a video picture that looks something like Figure 17-4. Your camera may have a Backlight setting to compensate for this condition, but it's best to shoot subjects against a more neutral or dark background whenever possible.
- **Wear sunscreen.** Your video image isn't the only thing you should protect from the sun!

Figure 17-4: Avoid backlit situations such as this.

Compositing Video

Moviemakers have had a lot of time, over the last century or so, to come up with creative techniques for making better movies. One of the most common special-effects techniques used today is called *compositing*. When you composite a scene in your video, you actually layer several different clips of video over each other to make a single image.

A common, everyday example of compositing can be seen during the weather forecast on the evening news. When you watch it on TV, it looks like the weatherperson is standing in front of a radar weather map, but in all likelihood, that map is actually a blank wall. Editing software superimposes the image of the weatherperson over the image of the weather map, creating the illusion you see on TV. (The weatherperson discreetly watches all this on a monitor screen so as not to lose track of where to point.)

You can do compositing too, if you have the right software. Unfortunately, it's not something that you can do *easily* with Apple iMovie or Pinnacle Studio, but slightly more advanced programs (such Adobe Premiere and Pinnacle Edition) can handle compositing just fine. The next couple of sections show you briefly how it works.

Understanding compositing

When your local weatherperson makes gestures that seem to help push a cold front across the continent on your screen, he or she is usually in front of a blank wall in the studio. When you watch it on TV, and you probably see a graphic weather map behind the weatherperson, a computer is creating the illusion. In effect, the computer treats the blank wall as transparent.

How does the computer know which parts of the video image to make transparent? Usually, that blank wall (or screen) behind the actor or subject is colored a specific, bright shade of blue or green. Computer software picks up on this unusual color and removes all occurrences of it from the video image, making the blue screen or green screen in the background disappear. (Of course, it's important to make sure that the actor's clothing doesn't have the same shade of blue or green, or that spot will disappear as well, and a piece of the background will show through the actor.)

Take a look at Figures 17-5, 17-6, and 17-7 to see how compositing looks. Figure 17-5 shows a video clip of an actor in front of a blue screen. Figure 17-6 shows the video clip that I'll use as a background. When the two clips are composited in video-editing software, the blue screen becomes transparent while the actor remains in view, as shown in Figure 17-7.

Figure 17-5: First, shoot the subject in front of a blue screen.

Figure 17-6: Next, choose a background clip.

Figure 17-7: The composited video image will look like this.

To remove the blue screen but not the subject, the editing software uses a technique called *keying*. A specific color is chosen as the *key color,* and the software removes the key color from the video image. Most programs use a tool called a *chroma key* for bluescreen or greenscreen effects.

Chroma isn't the shiny stuff that covers the massive bumpers on a '59 Cadillac. It's actually just a fancy word for color used by video pros, because I guess someone decided that "color key" didn't sound geeky enough.

Shooting video against a blue screen

Moviemakers use blue or green screens because they usually contrast with everything else in the video. Some professional moviemakers now feel that green screens work a little better than blue screens, but for most purposes, blue is probably good enough.

When you shoot video in front of a blue screen, the most important thing is to make sure that the screen is completely and evenly lit. This ensures that the entire screen shows up as a single shade, which allows the software to easily remove it later. In the next two sections, I show you how to build your own blue screen and shoot subjects in front of it.

Building a blue screen

Video equipment tends to be pretty expensive, so you'll be happy to learn that an effective blue screen can be built for less than $50. My favorite material for the purpose is blue plastic picnic-table covering (available at your local party-supply store) — it's cheap and effective. You can also use linen sheets (or even paint) of the correct color. Whatever material you use, make sure it meets the following requirements:

- **The material should have a matte finish.** If the finish is glossy, light will probably reflect off it in bright spots, creating uneven shades of blue.
- **The material should be available in bulk.** You'll want to cover an area much larger than the video scene you intend to shoot. (I like picnic-table

covering because it comes in big, cheap rolls.) Having a plentiful supply of material gives you some flexibility in positioning your subjects, lights, and camera.

I built the blue screen shown in this chapter by stapling the picnic-table material to a rudimentary frame built from 1x3 firring strips, available cheaply at any lumber store. It's important to pull the material so that it's flat and free of wrinkles. Surface imperfections show up as different shades of blue in the video image — which will prevent the blue color from being completely keyed out of the image.

Shooting subjects against the screen

For your composited video to be effective, it's absolutely critical that your blue screen be fully and evenly lit. A good start is to suspend a fluorescent light directly above the blue screen. This light should be between the subject and the blue screen so the subject doesn't create shadows. In addition, position a couple of halogen shop lights on the floor on either side of the scene, pointing at the blue screen. If these lights cause a glare on the blue screen, diffuse the light using a gel (as described in Chapter 4).

Lights that illuminate the subject should be diffused so the subject doesn't cast shadows on the blue screen. To diffuse light, bounce the light onto the subject via a reflector, or soften the light using a gel. Your blue screen set will probably look something like Figure 17-8.

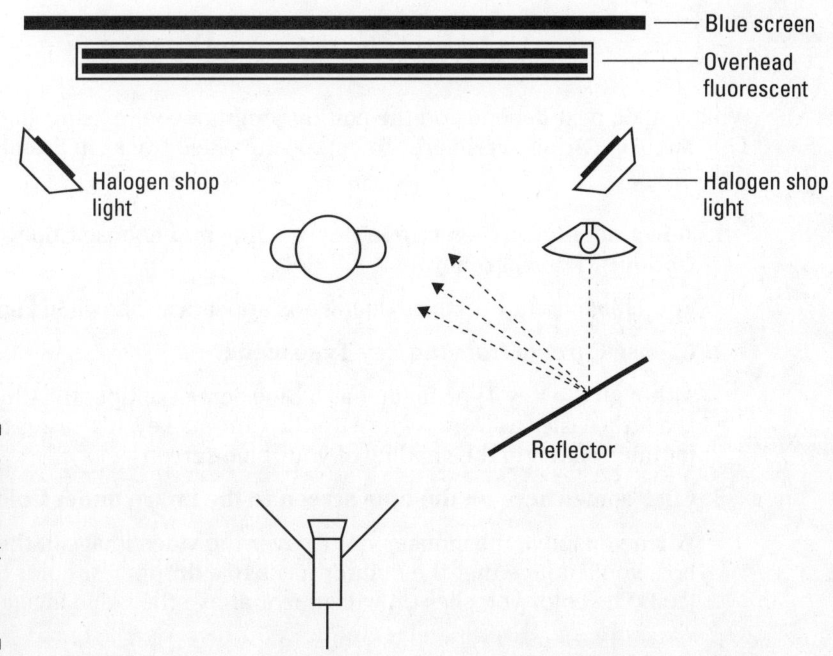

Figure 17-8: Make sure your blue screen is lit fully and evenly.

You will no doubt need to do some fine-tuning of the lighting. View it carefully, with a critical eye, to ensure that the entire screen has a uniform shade of blue. Also, make sure your subject isn't wearing the same shade of blue, unless you want to make it appear that there is a hole through the subject.

Compositing video tracks

Once your video is shot and imported into your editing program, compositing the video takes only a few steps. First, place the clip that will serve as the background in the first video track. Next, place the bluescreen clip in the track directly above it, as shown in Figure 17-9.

Figure 17-9: In Adobe Premiere, choose Blue Screen from the Key Type menu in the Transparency Settings dialog box.

What you do next depends on the editing program you're using. In Figure 17-9, I'm using Adobe Premiere. To composite video tracks in Premiere, follow these steps:

1. **Select the blue screen clip in the timeline and choose Clip ⇨ Video Options⇨Transparency.**

   The Transparency Settings dialog box appears as shown in Figure 17-9.

2. **Choose Chroma from the Key Type menu.**

   Although the Key Type menu has a Blue Screen option, the Chroma key is a little easier to work with. With the Chroma key, it's easier to correct for minor flaws in the shading of your blue screen.

3. **Click somewhere on the blue screen in the image under Color.**

   When you move the mouse pointer over the video image in the Color box, you'll notice that the pointer turns to a dropper or color picker icon. The color you click on will appear above the video image, and this is the color that the Chroma key will remove from the video image. The Sample image should show which parts of your image will now be transparent.

4. **If some parts of the blue screen still appear in the Sample image, adjust the Similarity and Blend sliders until the whole blue screen is invisible.**
5. **Click OK when you're done.**

The blue screen becomes invisible in the video image, making the composite effect complete!

In Adobe Premiere and many other video-editing programs, it may be necessary to render the timeline before you can preview your bluescreen effect. Rendering builds a preview file on your hard drive. To render the timeline in Premiere, choose Timeline⇨Render Work Area.

Chapter 18

Ten Tools for Digital Video Production

In This Chapter

- Tripods
- Video lights
- DVD burners
- Microphones
- MiniDisc recorders
- Multimedia controllers
- Video converters
- Graphical background video elements
- Helmet cams
- Video decks

If you've read much of this book, you know that I like to say things like, "All you need is a digital camcorder and a computer to make awesome movie productions!"

Sure, a camcorder and a computer may be all you *need* to make movies, but you can enhance your video projects with many other items. Some tools help you shoot better video or record better audio, and some tools can help you out when you edit your video.

Tripods and Other Stabilization Devices

The need for image stabilization will probably become apparent the first time you watch your footage on a large TV screen. No matter how carefully you try to hold the camera still, some movement is going to show up on the image. Of course, there are plenty of times when handheld is the way to shoot, but there are plenty of other times when a totally stable image is best. For that, you need a tripod.

Tripods are widely available, starting as low as $20 at your local department store. Alas, as with so many other things in life, when you buy a tripod you get what you pay for. High-quality video tripods incorporate several important features:

- **Dual-stanchion legs and bracing:** Dual-stanchion legs generally have two poles per leg, which gives the tripod greater stability, especially during panning shots. Braces at the base or middle of the tripod's legs also aid stability.
- **High-tech, lightweight materials:** You'll soon get tired of lugging a 15-20 pound tripod around with your camera gear. Higher-quality tripods frequently use high-tech materials (including titanium, aircraft-quality aluminum, and carbon-fiber) that are strong and lightweight, making the gear less cumbersome to transport and use.
- **Bubble levels:** These built-in tools help you ensure that your camera is level, even if the ground underneath the tripod isn't.
- **Fluid heads:** These ensure that pans will be smooth and jerk-free.
- **Counterweights:** The best tripods have adjustable counterweights so the head can be balanced for your camera and lens (telephoto lenses, for example, can make the camera a bit front-heavy). Counterweights allow smooth use of the fluid head while still giving you the option of letting go of the camera without having it tilt out of position.

For a tripod with all these features, you can expect to spend at least $300, if not much, much more. If that kind of money isn't in your tripod budget right now, try to get a tripod that incorporates as many of these features as possible.

Monopods

Tripods aren't the only stabilization devices available. You may also want to keep a monopod handy for certain occasions. As the name suggests, a monopod has only one leg (just as tripods have three legs, octopods have eight, and . . . never mind). Although this means some camera movement is inevitable — you have to keep the camera balanced on the monopod — resting the monopod on the ground can give you more stability than you'd have if you simply hand-held the camera. I used a monopod recently when I was shooting some video at a Taekwondo tournament in which my kids participated. The tournament floor was crowded with parents and competitors, so the wide footprint of a regular tripod would not have been practical. As a bonus, I could also use the monopod as a makeshift boom for overhead shots.

Mobile stabilizers

If you've watched a sporting event on TV recently, you may have seen footage of a player running along where the camera appears to stay with the player as he or she runs. And although the camera operator was obviously moving to get the shot, the image appears to be as stable as any tripod shot. How can this be? There are two possibilities:

- The camera operator is a superhero with the special ability to hold heavy objects absolutely still, even while riding a Radio Flyer wagon down a rocky, rutted slope.
- The camera operator was using a Steadicam-style device.

Steadicam (`www.steadicam.com`) is a brand of camera stabilizer that allows both super-stable images and handheld mobility. A Steadicam device attaches to the camera operator with an elaborate harness, and includes an LCD monitor that allows the operator to see the video image without taking her eyes off her path of travel. Steadicams are incredibly effective, but they are also incredibly expensive. The "affordable" Steadicam JR, which is aimed at semiprofessionals and prosumers, retails for $899! (You don't even want to know what the professional-grade units cost.)

Other, more affordable devices are available. Check your local camera shop to see what is available for mobile stabilization. If you're handy with a needle and thread, you can even make a sling to help stabilize your arm while shooting. Fashion a sling that enables you to support your forearm with your neck and shoulders. This will reduce fatigue and thus result in smoother images.

Lighting

Many video shoots can benefit from more light. Light brings out more detail and color in your video, which is why professional videographers always carefully set up lighting before shooting a scene. (Chapter 4 describes some techniques for lighting a video scene.) Lighting equipment that you'll want for your video shoots includes

- **Lights:** Most people would consider *lights* the most important pieces of lighting equipment. Incandescents, fluorescents, halogens, and most other bright light sources are useful.
- **Reflectors:** Reflectors bounce light onto a subject for a softer, more diffuse effect. You can buy professional-style reflectors, or you can make your own as I described in Chapter 4.

- **Gels:** Gels are translucent or semi-transparent materials that cover a light to diffuse it somewhat.
- **Extension cords:** Make sure you can get power to your lights.
- **Duct tape:** Use cheap, widely-available duct tape from your local hardware store, or be like the pros and pick up some theatrical gaffing tape from a musician-supply store. Tape is the miracle tool with a thousand-and-one uses. Use tape to hold up reflectors and gels. Secure cords to the floor so that they don't become trip hazards. Mark your coffee cup with a small strip of tape so you know which one is yours.

Professional-style video lighting can be purchased at most photographic supply stores, or you can mail order it from specialty retailers like B&H Photo (`www.bhphotovideo.com`). Be prepared, however, to spend hundreds (if not thousands) of dollars for professional equipment.

If you're more of a budget-minded moviemaker, do what I do and head down to the local hardware store. There you can find work lights that are bright, durable, and (most importantly) cheap. Work lights often come with stands or clips that help you position your lights without spending hours on setup. While you're at the hardware store, pick up some cheap fluorescent shop lights. These fixtures can be suspended over a scene or backdrop in your makeshift home studio and put out a lot of useful light for the price. Translucent fluorescent light covers make excellent, heat-resistant diffusers for light.

DVD-R Drives

When Apple first released its G4 Macintosh with SuperDrive in 2001, it seemed remarkable that a complete DVD (Digital Versatile Disc) authoring system could be had for *just* $5000. But prices dropped fast, and in less than a year, Apple was already selling iMacs capable of recording DVDs for less than $2000. Many new Macintosh computers can now be purchased with the SuperDrive (Apple's DVD burner), various new PCs have included DVD burners, and third-party DVD burners are now widely available as add-ons for existing Macintosh and PC systems. At this writing, DVD-R (DVD-Recordable) drives can be found for as little as $200, and I don't even want to speculate what they'll cost next week (or a few months from now, when you read this).

DVD burners — *burner* is the not-so technical term used interchangeably with *recorder* — are useful for a variety of reasons. First, they can record up to 4.7GB (gigabytes) of data onto a single disc. Second, you can record your movie projects directly onto movie DVDs so they can be watched in virtually any DVD player. (Chapter 16 describes the ins and outs of DVD recording.)

DVD-R drives can be purchased from a variety of sources. If you're planning to buy a new Macintosh soon, I recommend paying a little extra to make sure you get one with a SuperDrive. At this writing, Apple does not offer an accessory SuperDrive that you can add on to your current Mac. If you're buying a new Windows PC, just as with Macs, your life will be a lot easier if a DVD-R drive is pre-installed in the computer. But fortunately, many third-party alternatives are available from a variety of sources. Check your local computer retailer for options that may be available. DVD-R drives are available in two packaging flavors:

- **Internal:** These take up less desk space and cost less, but you'll need to have some computer hardware upgrading expertise to install the drive. Either that, or you could pay someone else to install the drive.
- **External:** Several different FireWire DVD-R drives are available. They cost a bit more but are much easier to install. You just connect the drive to your computer's FireWire port, install the software included with the drive, and you're ready to burn!

Microphones

Virtually all camcorders have built-in microphones. Most digital camcorders boast 48-bit stereo sound-recording capabilities, but you'll soon find that the quality of the audio you record is still limited primarily by the quality of the microphone you use. Therefore, if you care even a little about making great movies, you *need* better microphones than the one built into your camcorder.

Don't forget a slate!

If you use a secondary audio recorder, one of the biggest challenges you may face is synchronizing the audio it records with video. Professionals ensure synchronization of audio and video using a *slate* — that black-and-white board that you often see production people snapping shut on camera just before the director yells, "Action!"

The slate is not just a kitschy movie thing. The snapping of the slate makes a noise that can be picked up by all audio recorders on scene. When you are editing audio tracks later (see Chapter 10), this noise will show up as a visible spike on the audio waveform. Because the slate is snapped in front of the camera, you can later match the waveform spike on the audio track with the visual picture of the slate snapping closed on the video track. If you're recording audio with external recorders, consider making your own slate. This will make audio-video synchronization a lot easier during the post-production process.

Your camcorder should have connectors for external microphones, and your camcorder's manufacturer may offer accessory microphones for your specific camera.

One type of special microphone you may want to use is a *lavalier* microphone — a tiny unit that usually clips to a subject's clothing to pick up his or her voice. You often see lavalier mics clipped to the lapels of TV newscasters. Some lavalier units are designed to fit inside clothing or costumes, though some practice and special shielding may be required to eliminate rubbing noises.

A good place to look for high quality microphones is a musician's supply store. Just make sure that the connectors and frequency range are compatible with your camcorder or other recording device (check the documentation). You may also want to check with your camcorder's manufacturer; they might offer accessory microphones specially designed to work with your camcorder. Finally, the Internet is always a good resource as well. One good resource is `www.shure.com`, the Web site of Shure Incorporated. Shure sells microphones and other audio products, and the Web site is an excellent resource for general information about choosing and using microphones.

MiniDisc Recorders

Recording good sound used to mean spending hundreds of dollars for a DAT (digital audio tape) recorder. However, these days the best compromise of quality versus sound for an amateur moviemaker may be to use a MiniDisc recorder. MiniDisc player/recorders can record CD-quality audio onto MiniDiscs, and that audio can be easily imported into your movie project. Chapter 7 shows you how to import audio from CDs and MiniDiscs. Countless MiniDisc recorders are available for less than $200 from companies that include Aiwa, Sharp, and Sony.

I know, I know, your camcorder records audio along with video, and it's already perfectly synchronized. So what is the point of recording audio separately? Well, you may need the capabilities of an audio recorder in many situations. Take, for example, these three:

- **You may want to record a subject who is across the room.** In this case, have the subject hold a recorder (or conceal it so it's off camera), and attach an inconspicuous lavalier microphone to the subject.
- **You may want to record only a special sound, on location, and add it to the soundtrack later.** For example, you might show crashing waves in the distant background, but use the close-up sound of those waves for dramatic effect.

- **You can record narration for a video project.** After you've recorded your narration, tweak it till it suits your needs and then add it to the soundtrack of your movie.

Countless other uses for audio recorders exist. You may simply find that an external, dedicated recorder records better-quality audio than the built-in mic in your camcorder.

Multimedia Controllers

I don't know about you, but I find that manipulating some of the playback and editing controls in my video-editing software using just the mouse isn't always easy. Sure, there are keyboard shortcuts for most actions — and you may find yourself using those keyboard shortcuts quite a bit — but there's an even better way. You can also control most Windows or Macintosh programs with an external *multimedia controller,* such as the SpaceShuttle A/V from Contour A/V Solutions. The SpaceShuttle A/V, shown in Figure 18-1, features five buttons and a two-part, dial control in an ergonomically designed housing. The overall design of the SpaceShuttle A/V is based on professional video-editing controllers. The unit plugs into a USB port (which can be found on virtually any modern computer). You can find out more about the SpaceShuttle A/V online at

```
www.contouravs.com
```

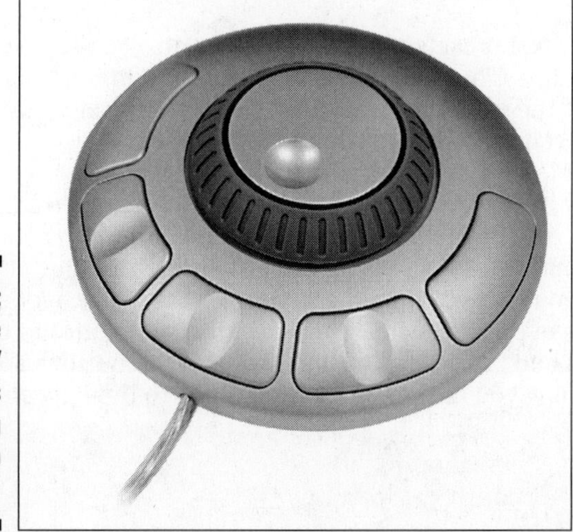

Figure 18-1: The Space Shuttle A/V makes editing video fun and easy!

Photo courtesy Contour A/V Solutions, Inc.

The SpaceShuttle A/V sells for about $50, and I tested it extensively with both Apple iMovie and Pinnacle Studio while writing this book. It works quite well with each program. If you decide to use a more advanced editing program such as Adobe Premiere or Apple Final Cut Express, you may want to consider the higher-end model, the ShuttlePro (shown in Figure 18-2). The ShuttlePro is similar to the SpaceShuttle A/V but with some additional buttons. The ShuttlePro also works well with iMovie and Studio, but many of the extra buttons are unused in these simpler programs.

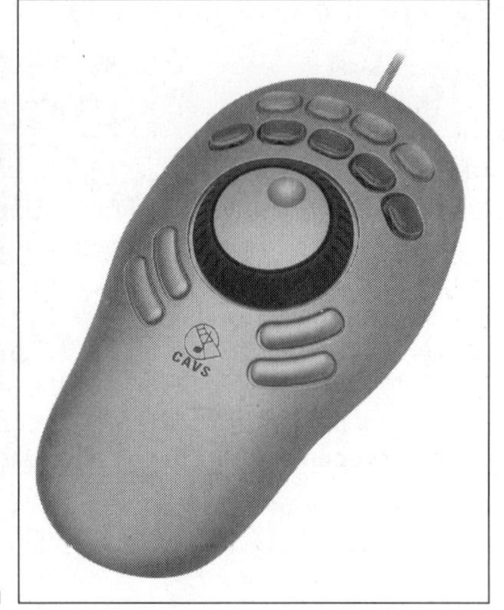

Figure 18-2: The ShuttlePro offers additional controls for more advanced video-editing programs.

Photo courtesy Contour A/V Solutions, Inc.

Contour provides control presets that allow their controllers to be used with many different programs. The array of buttons may look intimidating at first, but I have found that the various functions are learned quickly, and the ergonomic design certainly makes the controllers easier to use than keyboard shortcuts. In most video-editing programs, SpaceShuttle buttons can be used to play forward, play backward, pause, or move to edit points in a movie project. But most useful are the dials. The spring-loaded outer ring acts as a shuttle control, and the inner dial rolls video forward or back a frame at a time (like the jog control in many editing programs, only the dial is a *lot* easier to use). The controllers even work with device-control capabilities to control external camcorders and tape decks during capture. The buttons are customizable, too, meaning you can apply new functions to them if you want.

Video Converters

You have a computer with a FireWire port, and you want to capture some analog video. What are you going to do? You have many, many solutions, of course. You could install a video-capture card, but a good one is expensive and installing it means tearing apart your computer. If you're lucky, you might be able to connect an analog video source to the analog inputs on your digital camcorder, and *then* connect the camcorder to the FireWire port. This method is clumsy, however, and it simply won't work with some camcorders.

A simpler solution may be to use an external *video converter* — usually a box that connects to your computer's FireWire port. The box includes analog inputs, so you can connect an analog VCR or camcorder to the box. The unit itself converts signals from analog media into DV-format video, which is then captured into your computer — where you can easily edit it using your video-editing software.

If you have worked with analog video a lot, you're probably aware that some quality is lost every time you make a copy of the video (especially if it's a copy-of-a-copy). This is called *generational loss*. Not to worry: A video converter like those described here doesn't present any more generational loss than a standard video-capture card does; after the signal is converted to digital, generational loss is no longer a problem until you output the video to an analog tape.

Most converter boxes can also be useful for exporting video to an analog source. You simply export the DV-format video from your editing program, and the converter box converts it into an analog signal that you can record on your analog tape deck. Among other advantages, this method of export saves a lot of wear and tear on the tape-drive mechanisms in your expensive digital camcorder. Features to look for in a video converter include

- Analog output
- PC or Mac support (as appropriate)
- Color-bar output
- Multiple FireWire and analog inputs/outputs

Video converters typically range in price from $200 to $300 for consumer-level converters. Table 18-1 lists a few popular units.

Table 18-1 Video Converters

Manufacturer	*Model*	*Street Price*	*Web Site*
Canopus	ADVC-50	$200	`www.canopuscorp.com/`
Data Video	DAC-100	$200	`www.datavideo-tek.com/`
Dazzle	Hollywood DV Bridge	$225-250	`www.dazzle.com/`

Graphical Video Background Elements

If you're new to video editing, you probably find things like titles and simple transitions to be exceedingly cool. And they *are* cool, but eventually you may find that your projects need an extra graphical snap, so to speak. For example, you may find that black-and-white title screens are a little, well, *plain*.

A quick way to make title screens and other parts of your video more visually appealing is to put a fancy-looking image in the background. You could put a still picture behind the title (in fact, I show you exactly how to do this in Chapter 12), but hey, this is video we're talking about! Why not use a background that moves?

Several companies offer pre-made background elements that you can plop into your video projects. One such company is DV Logic (`www.dvlogic.com`), which offers a collection of 30 broadcast-quality elements for $99. Available elements include images of rippling waves, three-dimensional objects, swirling vortices, and many other designs. Figure 18-3 shows a rippling-wave element that is available for free download from the DV Logic Web site. I have placed the element in the iMovie timeline, and then created an overlay title (see Chapter 9 for more on working with titles) to use over it. The waves actually ripple in the background as the movie plays. Now *that* is cool!

Figure 18-3: Video elements can add a great deal of visual appeal to your projects.

Helmet Cams

If you participate in any high-action activities such as bicycling, motorcycling, skiing, or something similar, you probably can't wait to shoot some exciting video of your pastime. And why not? High-action activities are excellent video subjects, and one of the most interesting video shots you can get is one that shows the player's-eye-view.

Alas, it's really not prudent (or safe) to hold a camcorder as you ride your mountain bike down a narrow, rocky trail. A solution is to stow your camcorder in a backpack or other safe location and use a small external camera to shoot the video action. These small cameras are often called *helmet cams* because many people attach them to the side or top of a helmet. They connect to your camcorder with a cord and are small enough that you can safely mount them in a variety of locations. Several companies offer small accessory cameras:

- **bulletcam.com** at `www.helmet-cam.com`
- **CatchIt Cam** at `www.realandvirtual.com`
- **Cydonia Media Services** at `www.helmetcams.com`
- **SportZshot.com** at `www.sportzshot.com`
- **Viosport.com** at `www.viosport.com`

Usually the image quality captured by a small helmet cam is inferior to that recorded by your camcorder, but it's a small price to pay for great action shots. Before you choose any helmet cam, check the manufacturer's Web site to ensure that it will be compatible with your camcorder.

Although these small cams are often called "helmet" cams, I recommend you avoid attaching them to a helmet or any other place that may cause injury. And of course, protect your camcorder. The vibration and shaking (not to mention the danger of crashing) involved with riding a bicycle or skiing down a slope can damage your camcorder (among other things).

Video Decks

Because it's so easy to simply connect a FireWire cable to your camcorder and capture video right into your computer, you may be tempted to use your digital camcorder as your sole MiniDV tape deck. If you're on a really tight budget, you may not have much of a choice, but otherwise I strongly recommend a high-quality video deck. A video deck not only saves wear and tear on the tape drive mechanisms in your expensive camcorder, but it can also give you greater control over video capture and export back to tape.

Professional video decks are expensive, but if you do a lot of video editing, they quickly pay for themselves — both in terms of the greater satisfaction and quality you're likely to get from your finished movie, and in less money spent on camcorder maintenance. Table 18-2 lists some MiniDV decks to consider. The table also includes a Digital 8 video deck from Sony. If you have a Digital 8 camcorder you may want to consider one of these unique decks.

Table 18-2 DV Video Decks

Manufacturer	*Model*	*Formats*	*Retail Price*
JVC	HR-DVS3	MiniDV, S-VHS	$700-1000
Panasonic	AG-DV1000	MiniDV	$800-950
Sony	GVD-200 VCR Walkman	Digital 8	$600-700
Sony	GVD-1000 VCR Walkman with 4" LCD screen	MiniDV	$1100-1300

If you're shopping for a professional-grade MiniDV deck and money is no object, you can also look for decks that support the DVCAM or DVCPRO tape formats. These are more robust, professional-grade DV tape formats — and decks that support these formats generally also support MiniDV.

Even if you don't think a MiniDV deck is worth the expense, you may still want to consider a S-VHS deck. S-VHS (short for *Super*-VHS) is a higher-quality version of the venerable VHS videotape format. Many S-VHS decks are available for as little as $120 if you shop around. If you plan to export your movies back to tape, you'll get better recording quality if you export to an S-VHS deck instead of a conventional VHS deck. You can export from your computer directly to an S-VHS deck by using one of two methods:

- If you have an analog-video capture card such as the Pinnacle Studio Deluxe, or an analog-video converter like the Dazzle Hollywood DV Bridge, connect the S-VHS deck directly to the outputs for your analog device.
- Connect your digital camcorder to the FireWire port on your computer, and then connect the S-VHS deck to the analog outputs on the camcorder.

  Your camcorder acts as a converter, changing digital video from the computer into analog video for the S-VHS deck. Most modern digital camcorders allow you to do this, but a few older MiniDV and Digital8 camcorders may cause problems, so test your equipment to see what works. If your camcorder won't output analog audio at the same time as you export digital video from the computer, then you'll have to record the video onto the camcorder tape before exporting it to an analog video deck.

Chapter 19

Comparing Ten Video-Editing Programs

In This Chapter

- Adobe Premiere
- Apple iMovie, Final Cut Express, and Final Cut Pro
- Microsoft Windows Movie Maker
- Pinnacle Studio and Edition
- Roxio VideoWave
- Sonic Foundry Vegas
- Sonic Foundry Vegas and VideoFactory

When you browse the World Wide Web, it's more than likely that you do it using Microsoft Internet Explorer. And when you download technical documents or government forms from the Web, you probably use Adobe Acrobat to read them. These are just two examples of how a single program can come to dominate an entire genre of software.

Fortunately — or unfortunately, depending on how you look at it — there is no single dominating standard among video-editing programs, which are also sometimes called *NLE* (nonlinear editor) programs. You have many different programs to choose from, whether you're looking for something free or have a budget of thousands of dollars. This chapter helps you sort through and compare ten popular video editors (although this list should by no means be considered comprehensive). I'll run down a feature-by-feature comparison of the latest versions available as of this writing, so that you can determine which program might best fit your needs and budgets. Of course, free demos and trial versions are available for most video editors, and I recommend trying out any program you plan to buy, if at all possible. The ten programs compared in this chapter are

- Adobe Premiere
- Apple iMovie
- Apple Final Cut Express
- Apple Final Cut Pro
- Microsoft Windows Movie Maker
- Pinnacle Edition
- Pinnacle Studio
- Roxio VideoWave
- Sonic Foundry Vegas
- Sonic Foundry VideoFactory

Reviewing the Basics

In your review of video-editing programs, you need to start somewhere, and it might as well be with the basics. Table 19-1 helps you identify which programs will work on your computer, how much they cost, and where to find more information. Keep in mind that I have listed suggested retail prices in U.S. dollars, current as of this writing. Software vendors may change their prices in the future, or you might be able to take advantage of upgrade specials or discounted sale prices. Also, I only list prices for the basic software. Sometimes you can get software packaged with hardware — for example, Pinnacle sells capture cards that include the Studio or Edition software — but I don't list those packages here because there are too many possible combinations.

Before you purchase or install any software, make sure you check the specific system requirements for that program. Table 19-1 tells you whether each program works on a Mac, a Windows PC, or both, but beyond that, there are many other system requirements you should review. (See Chapter 2 for information on making your computer ready for editing video, regardless of which program you decide to use.)

Table 19-1 Basic Video Editors

Program	Retail Price	Platform	Web Address
Adobe Premiere	$549	Macintosh Windows	`www.adobe.com/products/premiere/`
Apple iMovie	Free (see Appendix C)	Macintosh	`www.apple.com/imovie/`

Program	Retail Price	Platform	Web Address
Apple Final Cut Express	$299	Macintosh	www.apple.com/finalcutexpress/
Apple Final Cut Pro	$999	Macintosh	www.apple.com/finalcutpro/
Microsoft Windows Movie Maker	Free (see Appendix E)	Windows	www.microsoft.com/windowsxp/moviemaker/
Pinnacle Edition	$699	Windows	www.pinnaclesys.com
Pinnacle Studio	$99	Windows	www.pinnaclesys.com
Roxio VideoWave (Power Edition)	$99	Windows	www.roxio.com/en/products/videowave_power_edition/
Sonic Foundry Vegas	$699	Windows	www.sonicfoundry.com
Sonic Foundry VideoFactory	$69	Windows	www.sonicfoundry.com

Importing Media

Moviemaking isn't much fun if you don't have some audio and video to work with. Virtually all editing programs can import directly from a DV (digital video) camcorder, and some can also import media in a variety of formats. Table 19-2 lists the formats in which these ten editing programs can capture. Some programs support additional formats not listed here, but the formats in Table 19-2 are the ones you're most likely to encounter. In Table 19-3, I have also noted which programs have a batch-capture feature (which, as mentioned in Chapter 11, can be a big timesaver when you capture video). Figure 19-1 shows Adobe Premiere's Batch Capture window.

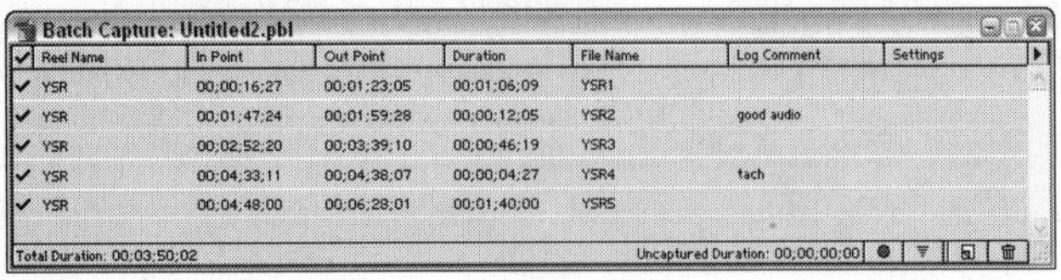

Figure 19-1: Batch-capture tools can be great timesavers.

Table 19-2 Supported Formats for Importing Media

Program	*AVI*	*CD Audio*	*MPEG*	*MP3*	*QuickTime*	*WindowsMedia*
Adobe Premiere	X	Mac only	X	X	X	
Apple iMovie		X (using iTunes)			X	
Apple Final Cut Express		X	X		X	
Apple Final Cut Pro		X	X		X	
Microsoft Windows Movie Maker	X	X (using Windows Media Player)	X	X		X
Pinnacle Edition	X	X	X	X	X	
Pinnacle Studio	X	X	X	X		
Roxio VideoWave	X	X	X	X		X
Sonic Foundry Vegas	X	X	X	X	X	X
Sonic Foundry VideoFactory	X		X	X	X	

Table 19-3 Support for Video Capture

Program	*DV*	*Analog*	*Batch Capture*
Adobe Premiere	X	X	X
Apple iMovie	X		
Apple Final Cut Express	X		
Apple Final Cut Pro	X	X	X
Microsoft Windows Movie Maker	X		
Pinnacle Edition	X	X	X
Pinnacle Studio	X	X	
Roxio VideoWave	X		
Sonic Foundry Vegas	X	X	X
Sonic Foundry VideoFactory	X		

Editing Your Movies

The primary differences between editing programs are the actual editing features offered by each one. Table 19-4 lists the number of video tracks, audio tracks, and transitions available in each program. Generally speaking, the more tracks you have to work with, the better. Figure 19-2 shows a timeline in Adobe Premiere that currently has three separate video tracks and three audio tracks — and more can be added if necessary. This arrangement provides a great deal of flexibility when you're editing. Table 19-4 also lists which programs offer tools to perform basic "cleanup" tasks on your video, such as improving color or lighting.

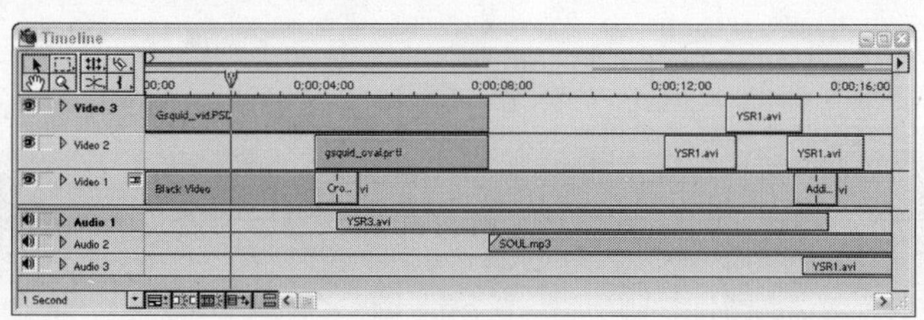

Figure 19-2: Multiple video and audio tracks give you more editing flexibility.

Table 19-4 Basic Editing Capabilities

Program	*Video Tracks*	*Audio Tracks*	*Transitions*	*Video Cleanup Tools*
Adobe Premiere	99	99	75	X
Apple iMovie	1	3	13	X
Apple Final Cut Express	99	99	60	X
Apple Final Cut Pro	99	99	77	X
Microsoft Windows Movie Maker	1	2	60	
Pinnacle Edition	Unlimited	Unlimited	200+	X
Pinnacle Studio			142	X
Roxio VideoWave	1*	6	60	X
Sonic Foundry Vegas	Unlimited	Unlimited	23	X
Sonic Foundry VideoFactory	2	3	174	X

**Roxio's VideoWave only has one real video track, but the Video Mixer feature allows you to composite two video clips as if you had two video tracks instead of one.*

Table 19-5 identifies more advanced editing features. Here you can see which programs allow you to animate objects on screen, control transparency of video clips, or key out certain parts of a video image using chroma keys or other similar tools. Keying is important if (for example) you want to do blue-screen effects.

Table 19-5 Advanced Editing Capabilities

Program	*Animation*	*Transparency*	*Keying*
Adobe Premiere	X	X	X
Apple iMovie			
Apple Final Cut Express	X	X	X
Apple Final Cut Pro	X	X	X
Microsoft Windows Movie Maker			

Program	*Animation*	*Transparency*	*Keying*
Pinnacle Edition	X	X	X
Pinnacle Studio			
Roxio VideoWave	X		X
Sonic Foundry Vegas	X	X	X
Sonic Foundry VideoFactory		X	

Most video-editing programs offer some built-in special effects. Table 19-6 shows you how many video effects and audio effects are offered in each program. You can also see which programs allow you to control effects using keyframes (which generally indicates far more advanced effects capabilities). Keyframes allow you to control an effect dynamically throughout a clip, as opposed to just applying it uniformly to an entire clip. Finally, some programs come with prerecorded sound effects that you can use in your movies. Interestingly, prerecorded sound effects are more common in the cheaper, low-end editing programs.

Table 19-6 Special-Effects Capabilities

Program	*Video Effects*	*Audio Effects*	*Keyframe Control*	*Sound Effects*
Adobe Premiere	79	24	X	
Apple iMovie	20	0		X
Apple Final Cut Express	75	12	X	
Apple Final Cut Pro	75+	25	X	X
Microsoft Windows Movie Maker	28	0		
Pinnacle Edition	Hundreds	Yes	X	
Pinnacle Studio	0	0		X
Roxio VideoWave	27	0		
Sonic Foundry Vegas	42	32	X	
Sonic Foundry VideoFactory	115	0		X

Table 19-7 lists some generators included with the various editing programs. Sorry, none of these generators will power your refrigerator during a brownout — what they do is produce elements that are useful in video production. A *bar-and-tone generator* produces color bars (see Figure 19-3) and a steady audio tone that, when recorded on tape, can be used to calibrate the colors and audio levels on broadcast and display equipment. Some programs also have a *counting leader* generator, which produces a visible, ten-second countdown for the beginning of the movie. Counting leaders also help professional video engineers calibrate their equipment, although a counting leader is basically just a novelty unless you actually produce video for broadcast.

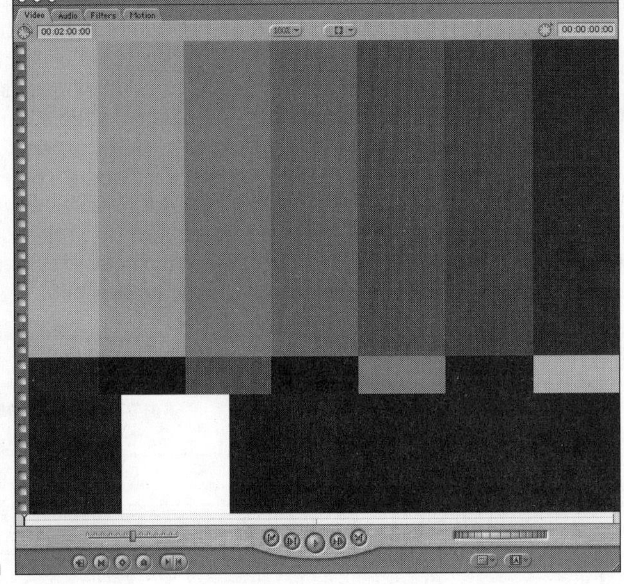

Figure 19-3: Color bars are calibrating broadcast and display equipment to work with your tape.

A *soundtrack generator,* on the other hand, is something we can all appreciate and use. Some video-editing programs (like Pinnacle Studio and Adobe Premiere) come with programs like SmartSound QuickTracks, which automatically generates high-quality soundtrack music in a variety of styles to match your project.

Table 19-7 Generators

Program	*Bars and Tone*	*Counting Leader*	*Soundtrack*
Adobe Premiere	X	X	X
Apple iMovie			

Program	*Bars and Tone*	*Counting Leader*	*Soundtrack*
Apple Final Cut Express	X	X	
Apple Final Cut Pro	X	X	X
Microsoft Windows Movie Maker			
Pinnacle Edition	X		
Pinnacle Studio			X
Roxio VideoWave			
Sonic Foundry Vegas	X		
Sonic Foundry VideoFactory			X

Exporting Your Movies

All video-editing programs provide tools to help you export your movie to the rest of the world. Table 19-8 shows you which popular export formats are available in each program. Some programs now come with tools to help you author DVDs as well — and that's noted in Table 19-8 as appropriate.

Table 19-8 Export Options

Program	*AVI*	*DVD*	*MPEG*	*QuickTime*	*RealMedia*	*WindowsMedia*
Adobe Premiere	X	Windows only	X	X	X	X
Apple iMovie		iDVD		X		
Apple Final Cut Express		iDVD	X	X		
Apple Final Cut Pro		iDVD	X	X		

(continued)

Table 19-8 *(continued)*

Program	*AVI*	*DVD*	*MPEG*	*QuickTime*	*RealMedia*	*WindowsMedia*
Microsoft Windows Movie Maker	X					X
Pinnacle Edition	X	X	X	X	X	X
Pinnacle Studio	X	X	X		X	X
Roxio VideoWave	X	X	X		X	X
Sonic Foundry Vegas	X	X	X	X	X	X
Sonic Foundry VideoFactory	X		X	X	X	X

Part VI
Appendixes

"It's a site devoted to the 'Limp Korn Chilies' rock group. It has videos of all their performances in concert halls, hotel rooms, and airport terminals."

In this part...

Sometimes, when you're making movies, you just need a quick reference to provide some essential bit of knowledge. I've provided several Appendixes to *Digital Video For Dummies* that serve as quick references on several subjects. Appendix A shows you how to take advantage of this book's companion CD-ROM. Appendix B is a glossary of the various technical terms and acronyms that seem to pop up everywhere when you work with video. The remaining appendixes help you get the most out of the three video-editing programs shown throughout this book: Apple iMovie, Pinnacle Studio, and Windows Movie Maker.

Appendix A

About the CD

In This Appendix

- Checking system requirements
- Using the CD with Windows and Mac
- Reviewing what you'll find on the CD
- Troubleshooting

System Requirements

Make sure your computer meets — or exceeds — the minimum system requirements shown in the following list. If your computer doesn't match up to most of these requirements, you may have problems using the software and files on the CD. For the latest and greatest information, please refer to the `ReadMe` file located at the root directory of the CD-ROM.

- A PC with a 500MHz AMD Athlon/Duron or Intel Pentium or faster processor; or a Mac OS computer with a 400MHz G3 or faster processor
- Microsoft Windows 98 Second Edition or later; or Mac OS system software 10.1.5 or later
- At least 128MB of total RAM installed on your computer (for best performance, we recommend at least 256MB)
- A CD-ROM drive
- A sound card for PCs (Mac-OS computers have built-in sound support)
- A monitor capable of displaying at least 16-bit color or higher; a Mac OS computer with a monitor capable of 1024 x 768 screen resolution or higher
- A modem with a speed of at least 14,400 bps

If you want to install the Pinnacle Studio trial software, you must have a Windows PC that meets the following additional requirements:

- 1000MHz (1.0GHz) or faster processor
- Windows XP recommended
- 300MB free hard-disk space for software installation; significant additional free drive space (about 1GB per minute of video) is required for video files and rendering or work space that will be used by the editing software as you edit
- Hard disk capable of read/write speeds of 4MB/second or better

If you need more information on the basics, check out these books published by Wiley Publishing, Inc.: *PCs For Dummies,* by Dan Gookin; *Macs For Dummies,* by David Pogue; *iMacs For Dummies,* by David Pogue; *Windows 98 For Dummies, Windows 2000 Professional For Dummies,* and *Microsoft Windows ME Millennium Edition For Dummies,* all by Andy Rathbone.

Using the CD with Microsoft Windows

To install from the CD to your hard drive, follow these steps:

1. **Insert the CD into your computer's CD-ROM drive.**
2. **Click the Start button and choose My Computer from the menu.**

   In Windows Me, double click the My Computer icon on your Windows desktop.
3. **Double Click the icon for your computer's CD-ROM drive.**
4. **Double click the file called License.txt.**

   This file contains the end-user license that you agree to by using the CD. When you are done reading the license, close the program, most likely NotePad, that displayed the file.
5. **Double click the file called Readme.txt.**

   This file contains instructions about installing the software from this CD. It might be helpful to leave this text file open while you are using the CD.
6. **Double click the folder for the software you are interested in.**

   Be sure to read the descriptions of the programs in the next section of this appendix (much of this information also shows up in the Readme file). These descriptions will give you more precise information about the programs' folder names, and about finding and running the installer program.

7. **Find the file called Setup.exe, or Install.exe, or something similar, and double click on that file.**

   The program's installer will walk you through the process of setting up your new software.

To run some of the programs, you may need to keep the CD inside your CD-ROM drive. This is a Good Thing. Otherwise, the installed program would have required you to install a very large chunk of the program to your hard drive space, which would have kept you from installing other software.

Using the CD with Mac OS

To install the items from the CD to your hard drive, follow these steps.

1. **Insert the CD into your computer's CD-ROM drive.**

   In a moment, an icon representing the CD you just inserted appears on your Mac desktop. Chances are, the icon looks like a CD-ROM.

2. **Double click the CD icon to show the CD's contents.**
3. **Double click the file called License.txt.**

   This file contains the end-user license that you agree to by using the CD. When you are done reading the license, you can close the window that displayed the file.

4. **Double click the Read Me First icon.**

   This text file contains information about the CD's programs and any last-minute instructions you need to know about installing the programs on the CD that we don't cover in this appendix.

5. **Most programs come with installers—with those you simply open the program's folder on the CD and double click the icon with the words "Install" or "Installer," or sometimes "sea" (for self-extracting archive).**
6. **If you don't find an installer, just drag the program's folder from the CD window and drop it on your hard drive icon.**

After you have installed the programs that you want, you can eject the CD. Carefully place it back in the plastic jacket of the book for safekeeping.

What You'll Find on the CD

The following sections are arranged by category and provide a summary of the software and other goodies you'll find on the CD. If you need help with installing the items provided on the CD, refer back to the installation instructions in the preceding section.

Shareware programs are fully functional, free, trial versions of copyrighted programs. If you like particular programs, register with their authors for a nominal fee and receive licenses, enhanced versions, and technical support. *Freeware programs* are free, copyrighted games, applications, and utilities. You can copy them to as many PCs as you like — for free — but they offer no technical support. *GNU software* is governed by its own license, which is included inside the folder of the GNU software. There are no restrictions on distribution of GNU software. See the GNU license file in the root directory of the CD for more details. *Trial, demo,* or *evaluation* versions of software are usually limited either by time or functionality (such as not letting you save a project after you create it).

Sample video clips

For Windows and Mac.

If you don't have your own video to work with yet, you can use the sample video clips in the Samples directory on the CD. Clips are included for both Macintosh and Windows computers. To access a clip, click the Samples folder, and then click the folder for the chapter whose sample clips you would like to access.

Pinnacle Studio

Trial version.

For Windows. Pinnacle Studio is a video-editing program that provides a powerful set of tools for a very reasonable price. The trial software is fully functional for 30 days, after which you can purchase the commercial version from a variety of retailers or directly from Pinnacle for $99.99. Studio is featured in examples throughout this book.

For more information or to purchase the software, visit

```
www.pinnaclesys.com/studio8democd/
```

Links Page

The links page contains hyperlinks to and brief descriptions of the many helpful Web sites discussed in this book.

Troubleshooting

I tried my best to compile programs that work on most computers with the minimum system requirements. Alas, your computer may differ, and some programs may not work properly for some reason.

The two likeliest problems are not having enough memory (RAM) for the program you want to use, or having other programs running that are affecting the installation (or running) of a program. If you get an error message such as `Not enough memory` or `Setup cannot continue`, try one or more of the following suggestions before you try using the software again:

- **Turn off any antivirus software running on your computer.** Installation programs sometimes mimic virus activity and may make your computer incorrectly believe that it's being infected by a virus.
- **Close all running programs.** The more programs you have running, the less memory is available to other programs. Installation programs typically update files and programs; so if you keep other programs running, installation may not work properly.
- **Have your local computer store add more RAM to your computer.** This is, admittedly, a drastic and somewhat expensive step. However, if you have a Windows PC or a Mac OS computer, adding more memory can really help the speed of your computer and allow more programs to run at the same time. Such an approach may include closing the CD interface and running a product's installation program from Windows Explorer.

If you still have trouble with the CD, please call the Customer Care phone number: (800) 762-2974. Outside the United States, call 1-(317)-572-3994. You can also contact Customer Service by e-mail at `techsupdum@wiley.com`. Wiley Publishing, Inc. will provide technical support only for installation and other general quality-control items; for technical support on the applications themselves, consult the program's vendor or author.

Appendix B

Glossary

alpha channel: Some digital images have transparent areas; the transparency is defined using an alpha channel.

analog: Data recorded as a wave with infinitely varying values is analog data. Analog recordings are usually electromechanical, so they often suffer from generational loss. *See also digital, generational loss.*

aspect ratio: The shape of a video image (its width compared to height) is the *aspect ratio*. Traditional television screens have an aspect ratio of 4:3, meaning the screen is four units wide and three units high. Some newer HDTVs use a "widescreen" aspect ratio of 16:9. Image pixels can also have various aspect ratios. *See also HDTV, pixel.*

bars and tone: A video image that serves the function of the "test pattern" used in TV broadcasting: Standardized color bars and a 1-kHz tone are usually placed at the beginning of video programs. This helps broadcast engineers calibrate video equipment to the color and audio levels of a video program. The format for color bars is standardized by the SMPTE. *See also SMPTE.*

bit depth: The amount of data that a single piece of information can hold depends upon how many bits are available. Bit depth usually measures color or sound quality. A larger bit depth means a greater range of color or sound.

black and code: The process of recording black video and timecode onto a new camcorder tape. This helps prevent timecode breaks. *See also timecode, timecode break.*

Bluetooth: A wireless networking technology that allows devices to connect to the Internet (or to computers) using radio waves. Some newer digital camcorders incorporate Bluetooth technology, although its practical applications for digital video have not yet been realized.

capture: The process of recording digital video or other media from a camcorder or VCR tape onto a computer's hard disk.

CCD (charged coupled device): This is the unit in camcorders that interprets light photons and converts the information into an electronic video signal. This signal can then be recorded on tape. Digital still cameras also use CCDs.

chrominance: A fancy word for color. *See also luminance.*

clip: One of various segments making up the scenes of a video program. Individual clips are edited into your video-editing program's timeline to form complete scenes and a complete story line. *See also storyboard, timeline.*

coaxial: Most wires that carry a cable TV signal use coaxial connectors. Coaxial connectors are round in cross-section, and have a single thin pin running through the middle of the connector. Coaxial cables carry both sound and video. Most TVs and VCRs have coaxial connectors; digital camcorders usually do not. Coaxial cables usually provide inferior video quality to component, composite, and S-Video cables. *See also component video, composite video, S-Video.*

codec: A scheme used to compress, and later decompress, video and audio information so it can pass more efficiently over computer cables and Internet connections to hard drives and other components.

color gamut: The total range of colors a given system can create (by combining several basic colors) to display a video image. The total number of individual colors that are available is finite. If a color cannot be displayed correctly, it is considered *out of gamut.*

color space: The method used to generate color in a video display. *See also color gamut, RGB, YUV.*

component video: A high-quality connection type for analog video. Component video splits the video signal and sends it over three separate cables, usually color-coded red, green, and blue. Component video connections are unusual in consumer-grade video equipment, but they provide superior video quality to coaxial, composite and S-Video connections. *See also analog, coaxial, composite video, S-Video.*

composite video: A connection type for analog video, typically using a single video-connector cable (color-coded yellow). The connector type is also sometimes called an *RCA connector*, usually paired with red and white audio connectors. Composite video signals are inferior to S-Video or component video because they tend to allow more signal noise and artifacts in the video signal. *See also analog, coaxial, component video, S-Video.*

DAT (digital audio tape): A digital tape format often used in audio recorders by professional video producers.

data rate: The amount of data that can pass over a connection in a second while contained in a signal. The data rate of DV-format video is 3.6MB (megabytes) per second.

device control: A technology that allows a computer to control the playback functions on a digital camcorder (such as play, stop, and rewind). Clicking Rewind in the program window on the computer causes the camcorder tape to actually rewind.

digital: A method of recording sound and light by converting them into data made up of discrete, binary values (expressed as ones and zeros). *See also analog.*

Digital 8: A digital camcorder format that uses Hi8 tapes. *See also digital, DV, MicroMV, MiniDV.*

DIMM (Dual Inline Memory Module): A memory module for a Mac or PC. Most computer RAM today comes on easily replaced DIMM cards. *See also RAM.*

driver: Pre-1980, the person in control of a car or horse-drawn carriage. Post-1980, a piece of software that allows a computer to utilize a piece of hardware, such as a video card or a printer.

drop-frame timecode: A type of timecode specified by the NTSC video standard, usually with a frame rate of 29.97fps. To maintain continuity, two frames are dropped at the beginning of each minute, except for every tenth minute. *See also timecode.*

DV (Digital Video): A standard format and codec for digital video. Digital camcorders that include a FireWire interface usually record DV-format video. *See also codec, FireWire.*

DVCAM: A professional-grade version of the MiniDV digital-tape format developed by Sony. DVCAM camcorders are usually pretty expensive. *See also digital, DVCPro, MiniDV.*

DVCPro: A professional-grade version of the MiniDV digital-tape format developed by Panasonic. Like DVCAM camcorders, DVCPro camcorders are usually very expensive. *See also digital, DVCAM, MiniDV.*

DVD (Digital Versatile Disc): A category of disc formats that allows capacities from 4.7 GB up to 17 GB. DVDs are quickly becoming the most popular format for distributing movies. Recordable DVDs (DVD-Rs) are becoming a common and affordable recording medium for home users.

EDL (Edit Decision List): A file or list that contains information about all edits performed in a program. This list can then be used to reproduce the same edits on another system, such as at a professional video-editing facility. Most advanced editing programs can generate EDLs automatically.

EIDE (Enhanced Integrated Drive Electronics): Most modern PCs and Macs have hard disks that connect to the computer using an EIDE interface. For digital video, you should try to use EIDE disks with a speed of 7200rpm.

exposure: The amount of light allowed through a camcorder's lens. Exposure is controlled with a part of the camcorder called the iris.

field: One of two separate sets of scan lines in an interlaced video frame. Each field contains every other horizontal resolution line. Immediately after one field is drawn on the screen, the other is drawn in a separate pass while the previous frame is still glowing, resulting in a complete image. *See also frame.*

FireWire: Also known by its official designation IEEE-1394, or by other names such as i.Link, FireWire is a high-speed computer peripheral interface standard developed by Apple Computer. FireWire is often used to connect digital camcorders, external hard disks, and some other devices to a computer. The speed of FireWire has contributed greatly to the affordability of modern video editing.

frame: Still image, one in a sequence of many that make up a moving picture. *See also frame rate.*

frame rate: The speed at which the frames in a moving picture change. Video images usually display 25 to 30 frames per second, providing the illusion of movement to the human eye. Slower frame rates save storage space, but can produce jerky motion; faster frame rates produce smoother motion but have to use more of the recording medium to store and present the images.

gamut: *See color gamut.*

gel: A translucent or colored sheet of plastic that is placed in front of a light to diffuse the light or change its appearance.

generational loss: A worsening of the signal-to-noise ratio (less signal, more noise) that occurs every time an analog recording is copied; some values are lost in the copying process. Each copy (especially a copy of a copy) represents a later, lower-quality *generation* of the original. *See also analog.*

HDTV (High-Definition Television): A new set of broadcast-video standards that incorporate resolutions and frame rates higher than those used for traditional analog video. *See also NTSC, PAL, SECAM.*

IEEE-1394: *See FireWire.*

i.Link: *See FireWire.*

interlacing: Producing an image by alternating sets of scan lines on-screen. Most video images are actually composed of two separate fields, drawn on consecutive passes of the electron gun in the video tube. Each field contains every other horizontal resolution line of a video image. Each field is drawn so quickly that the human eye perceives a complete image. *See also progressive scan, field.*

iris: *See exposure.*

jog: *See scrub.*

key light: A light used to directly illuminate a subject in a video shot.

lavalier: A tiny microphone designed to clip to a subject's clothing. Lavalier mics are often clipped to the lapels of TV newscasters.

lens flare: A light point or artifact that appears in a video image when the sun or other bright light source reflects on the lens.

luminance: A fancy word for brightness in video images. *See also chrominance.*

MicroMV: A small digital-camcorder tape format developed by Sony for ultra-compact camcorders. *See also digital, Digital 8, DV, MiniDV.*

MiniDV: The most common tape format used by digital camcorders. *See also digital, Digital 8, DV, MicroMV.*

moiré pattern: A wavy or shimmering artifact that appears in video images when tight parallel lines appear in the image. This problem often occurs when a subject wears a pinstriped suit or coarse corduroy.

NLE (nonlinear editor): A computer program that can edit video, audio, or other multimedia information without confining the user to an unchangeable sequence of frames from first to last. Using an NLE, you can edit the work in any order you choose. Popular video NLEs include Apple iMovie, Pinnacle Studio, and Windows Movie Maker.

NTSC (National Television Standards Committee): The broadcast-video standard used in North America, Japan, the Philippines, and elsewhere. *See also PAL, SECAM.*

online/offline editing: When you edit using full-quality footage, you are performing *online* editing. If you perform edits using lower-quality captures, and intend to apply those edits to the full-quality footage later, you are performing *offline* editing.

overscan: What happens when a TV cuts off portions of the video image at the edges of the screen. Most standard TVs overscan to some extent.

PAL (Phase Alternating Line): The broadcast-video standard used in Western Europe, Australia, Southeast Asia, South America, and elsewhere. *See also NTSC, SECAM.*

PCI: Peripheral Component Interconnect, a standard type of connection for computer expansion cards (such as FireWire cards). Any new card must be placed in an empty PCI slot on the computer's motherboard.

pixel: The smallest element of a video image, also called a *picture element.* Bitmapped still images are made up of grids containing thousands, even millions of pixels. A screen or image size that has a resolution of 640 x 480 is 640 pixels wide by 480 pixels high.

Plug and Play: A hardware technology that allows you to easily connected devices such as digital camcorders to your computer. The computer automatically detects the device when it is connected and turned on.

progressive scan: A scan display that draws all the horizontal resolution lines in a single pass. Most computer monitors use progressive scan. *See also interlacing.*

RAM (Random-Access Memory): The electronic working space for your computer's processor and software. To use digital video on your computer, you need lots of RAM.

render: To produce a playable version of an altered video image. If an effect, speed change, or transition is applied to a video image, your video-editing program must figure out how each frame of the image should look after the change. Rendering is the process of applying these changes. Usually, the rendering process generates a preview file that is stored on the hard disk. *See also transition.*

resolution: *See pixel.*

RGB (Red-Green-Blue): The color space (method of creating on-screen colors) used in computer monitors; all the available colors result from combining red, green, and blue pixels. *See also color space, YUV.*

sampling rate: The number of samples obtained per second during a digital audio recording. When audio is recorded digitally, the sound is sampled thousands of times per second. 48 kHz audio has 48,000 samples per second.

scrub: To move back and forth through a video program, one frame at a time. Some video-editing programs have a scrub bar located underneath the video preview window (also called the *jog control*). *See also shuttle.*

SECAM (Sequential Couleur Avec Memoire): Broadcast video standard used in France, Russia, Eastern Europe, Central Asia, and elsewhere. *See also NTSC, PAL.*

shuttle: To roll a video image slowly forward or back, often to check a detail of motion. Professional video decks and cameras often have shuttle controls. Some video-editing programs also have shuttle controls in their capture windows. *See also scrub.*

slate: The black-and-white hinged board that moviemakers snap closed in front of the camera just before action commences. The noise made by the snapping slate is used later to synchronize video with sound recorded by other audio recorders during the shoot.

SMPTE (Society for Motion Picture and Television Engineers): This organization develops standards for professional broadcasting equipment and formats. Among other things, the SMPTE defines standards for bars and tone, counting leaders, and timecode.

storyboard: A visual layout of the basic scenes in a movie. The storyboard can be used to arrange clips in a basic order before you do finer edits using the timeline. *See also clip, timeline.*

S-VHS: A higher-quality version of the VHS videotape format. S-VHS VCRs usually have S-Video connectors. *See also S-Video.*

S-Video: A high-quality connection technology for analog video. S-Video connectors separate the color and brightness signals, resulting in less signal noise and fewer artifacts. Most digital camcorders include S-Video connectors for analog output. Analog capture cards and S-VHS VCRs usually have S-Video connectors as well. *See also analog, capture, coaxial, component video, composite video, S-VHS.*

timecode: The standard system for identifying individual frames in a movie or video program. Timecode is expressed as *hours:minutes:seconds:frames* (as in 01:20:31:02). This format has been standardized by the SMPTE. Non-drop-frame timecode uses colons between the numbers; drop-frame timecode uses semicolons. *See also SMPTE, timecode break.*

timecode break: An inconsistency in the timecode on a camcorder tape. *See also timecode.*

timeline: The working space in most video-editing programs. Clips are arranged along a timeline, which may include different video tracks, audio tracks, or other features. The timeline usually shows more detail and allows finer editing control than the storyboard. *See also clip, storyboard.*

title: Text that appears on-screen to display the name of the movie, or to give credit to the people who made the movie. *Subtitles* are a special type of title, often used during a video program to show translations of dialogue spoken in foreign languages.

transition: The method by which one clip ends and another begins in a video program. A common type of transition is when one clip gradually fades out as the next clip fades in. *See also clip, render.*

USB (Universal Serial Bus): This is a computer port technology that makes it easy to connect a mouse, printer, or other device to a computer. Although USB usually isn't fast enough for digital-video capture, some digital camcorders have USB ports. Connected to a computer's USB port, these cameras can often be used as Webcams. Most computers built after Spring 2002 use a newer, faster version of USB called USB 2.0.

video card: This term can refer to either of two different kinds of devices inside a computer: the device that generates a video signal for the computer's monitor, or the card that captures video from VCRs and camcorders onto the computer's hard disk. Some hardware manufacturers refer to their FireWire cards as video cards because FireWire cards are most often used to capture video from digital camcorders. *See also capture, FireWire.*

waveform: A visual representation of an audio signal. Viewing a waveform on a computer screen allows precise synchronization of sound and video.

YUV: The acronym for the color space used by most TVs. For some reason YUV stands for *luminance-chrominance. See also chrominance, color space, luminance, RGB.*

zebra pattern: An overexposure-warning feature that some high-end camcorders have. A striped pattern appears in the viewfinder over areas of the image that will be overexposed unless the camcorder is adjusted to compensate.

Appendix C

Installing Apple iMovie

In This Appendix

- Installing iMovie
- Enhancing iMovie with plug-ins

One of the programs I feature in this book is Apple iMovie. If you're using a Mac, this is easily the best free video-editing software available. Unfortunately, as I write this, iMovie is also the *only* free editing software that is widely available for your Mac. It's a shame that more third-party software vendors aren't providing affordable editing programs for the Macintosh operating system — because Macs really do make excellent video-editing machines. One new alternative that should be available as you read this is Avid Free DV (`www.avid.com/avidfreedv`), which will be both free and available for OS X.

This appendix shows you briefly how to obtain and install the latest version of iMovie on your computer. I'll also show you how to find and install plug-ins to expand iMovie's capabilities.

Installing and Upgrading iMovie

The video-editing instructions in this book assume you have iMovie 3 or later. If you've just recently purchased a new Mac, you probably already have such a version. Version 3 of iMovie incorporated some important improvements, among them the capability to import QuickTime media clips into the program. If you have iMovie 1 or 2, you won't be able to import and use the sample clips on the CD-ROM that accompanies this book. iMovie also comes with iLife, a suite of software that includes current versions of iTunes, iPhoto, and iDVD.

To check your version of iMovie, launch the program from your applications folder and choose iMovie⇨About iMovie. A splash screen appears (as shown in Figure C-1), showing you the current version number.

Figure C-1: Make sure you have the latest version of iMovie.

To update iMovie, or just to check and see whether an update is available, do one of the following:

- Visit www.apple.com/imovie and click one of the download links on that page.
- If Software Update runs automatically on your system, check for iMovie updates when it runs.
- To run Software Update manually, open the System Preferences window (Apple⇨System Preferences) and open the Software Update icon. Click the Update Now button in the Software Update settings dialog box that appears.

When you download an update, you'll see a desktop icon for an iMovie installer disk image. Double-click that icon to temporarily mount a disk image for the installer, and then double-click the icon for the mounted disk image. A window will open with an installer package icon (Figure C-2). Yep, now you have to double-click that too. I know, it's a whole lotta icons to double-click, but it's actually a pretty simple process.

When you double-click the installer package, the first window you'll probably see is an Authorization dialog box. If you see a message stating that you must enter an administrator password, click the little lock icon at the bottom of the dialog box and enter an administrator password. Once that's done, just follow the instructions on-screen to install your update.

When you're done installing the software, you can get rid of the mounted-disk image by dragging it to the Trash bin. You can also delete the disk image file if you like, although I recommend backing up your update files in a safe place (for example, on a recordable CD).

Sometimes when you install a newer version of iMovie, you may also be required to install a new version of QuickTime or other software as well. Usually the Apple Web site offers recommendations on other software to download. Also, when you launch your updated version of iMovie you may see warnings about required software that isn't installed.

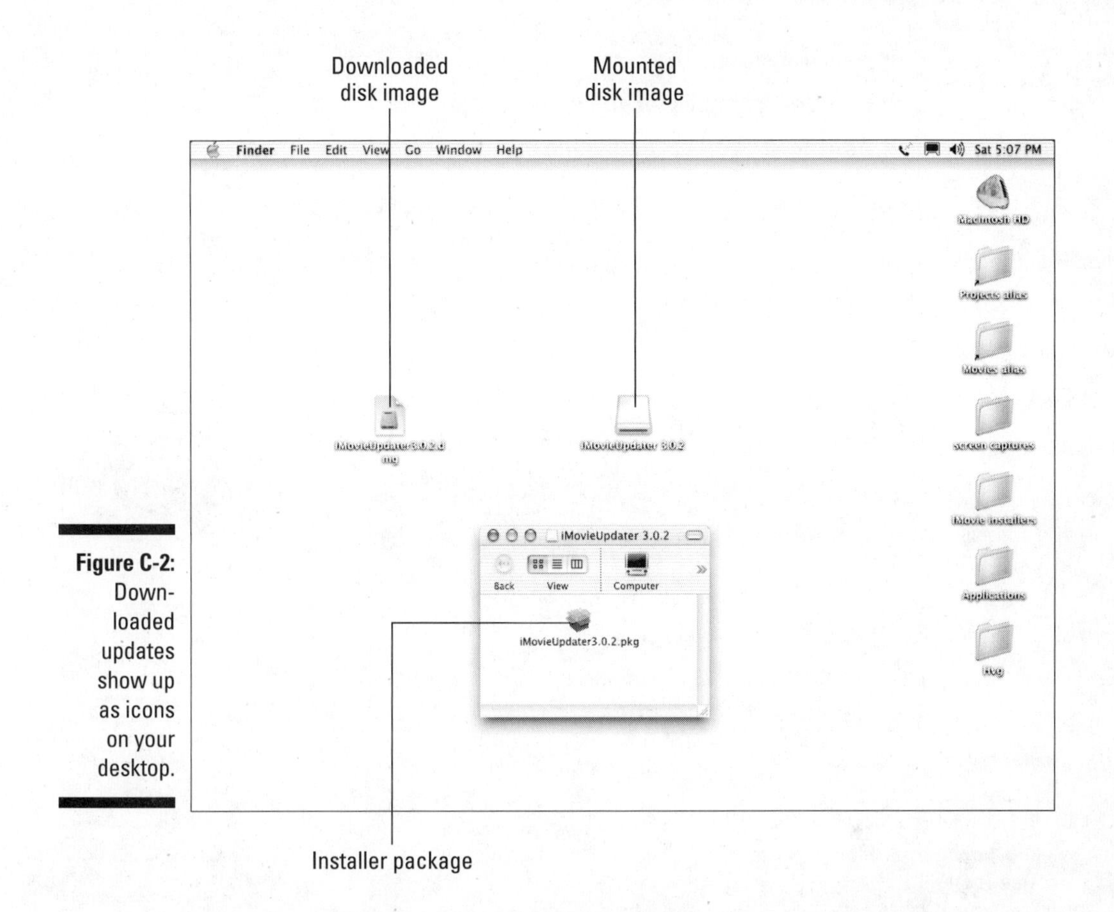

Figure C-2: Downloaded updates show up as icons on your desktop.

I strongly recommend that you use Software Update in the OS X Applications folder to download required software updates, especially QuickTime. I have found that not all parts of the Apple Web site talk to each other — meaning that links on Web sites don't always lead to the latest and greatest version. Fortunately, Apple does a pretty good job of keeping the available downloads in Software Update, er, *updated*, so I recommend using it first.

Enhancing iMovie with Plug-ins

The Apple Web site is a pretty useful resource for all things iMovie. If you aren't already there, go ahead and pay a visit to `www.apple.com/imovie`. Among other things you'll usually find cool audio and video effects that you can download and install as plug-ins to iMovie. Plug-ins are developed by Apple and third-party software vendors, and they may add a variety of capabilities to iMovie, including

- New transitions
- Borders or frames for the video image
- Advanced overlays and video effects

The current selection of plug-ins for iMovie may vary, but scroll down the iMovie Web page and look for links to downloadable plug-ins. Figure C-3 shows an X-Ray effect that that comes on an effects sampler from Gee Three (`www.geethree.com`).

Figure C-3: This X-Ray effect from Gee Three gives my video image an eerie look.

Appendix D

Using Pinnacle Studio

In This Appendix

- Updating Pinnacle Studio
- Installing and using Studio plug-ins

As I showed in Chapter 19, there is a pretty big selection of free or low-cost video-editing programs available for Windows PCs. Covering all of these programs completely would be impossible in a single book, so this book has focused on Pinnacle Studio. I chose Studio because it is widely available and offers an excellent balance of features for the price. Sure, you can download Microsoft Windows Movie Maker for free, but Studio offers a greater selection of export formats and offers far better quality, which is important if you intend to record your edited movies on videotape or DVD.

The companion CD-ROM for this book includes a 30-day free trial of the Studio software, and you can buy the full version of Studio at most electronics and computer retailers. Pinnacle Studio has a retail price of $99, but if you shop around, you can often find the software on sale for a little less. Pinnacle also sells FireWire and analog-video capture cards packaged with the Studio software for $129 (suggested retail). Although most FireWire cards come with some type of editing software, I think it's worth spending a few dollars on the Pinnacle card that comes with Studio because the program is more capable than most other free and low-cost editors. The ultimate Studio package is Studio Deluxe, which includes the Studio editing software, video-capture hardware for both digital and analog video, and the Hollywood FX Plug-in package, which provides some advanced video transitions and editing capabilities. Pinnacle Studio Deluxe retails for $299.

This appendix shows you how to keep your Studio software updated, as well as how to find and install useful plug-ins.

Updating Studio

Pinnacle frequently provides updates for its hardware and software online, and I recommend checking for updates on a regular basis (I usually do so at the beginning of each month). You should also check for online updates immediately after installing the software for the first time. There's a good chance that some updates have been made since your install disc was manufactured.

Checking for updates to the Studio software is really quite simple. Launch Studio and then choose Help⇨Software Updates. (You must have an active Internet connection on your computer before the software checks for updates online.) Studio automatically checks the Pinnacle Web site, and if any updates are available, it gives you instructions on how to download and install those updates.

If you're unable to open your Studio software, visit the Pinnacle Web site at `www.pinnaclesys.com` and click the Support link. When a list of products appears, click the Pinnacle Studio link. This will take you to a Web page with special information for Studio users, downloads, software updates, camcorder-compatibility lists, and more.

In addition to Studio, you should also regularly update

- **Hardware drivers:** If you have a Pinnacle capture card, check for updated drivers or hardware installers on a regular basis. *Drivers* are software tools that tell Windows how to use a piece of hardware such as a capture card. When I first installed my Pinnacle AV/DV capture card, it would frequently cause Windows XP to freeze during analog-video capture. I was about to pull my hair out in frustration, but instead I downloaded and ran a newer hardware installer from the Pinnacle Web site. Not only did this solve the freezing problem, but I was able to keep my hair.
- **Video-card drivers:** The video card is the component in your computer that generates the video picture that appears on your monitor. Outdated drivers cause a surprising number of problems in Windows, especially when you try to edit video. This seems to be especially true with some NVIDIA video cards, which frequently benefit from driver updates. If you're not sure what kind of video card you have, open the Windows Control Panel (Start⇨Control Panel), click Performance and Maintenance (if you see it), and open the System icon. On the Hardware tab, click Device Manager. Click the plus-sign next to Display Adapters to see the brand and model of your video card. Updated drives for NVIDIA cards can be downloaded from `www.nvidia.com`. If a different company made your video card, do a Web search on the company's name to locate the correct Web site.
- **Windows:** By default, Windows XP checks for system software updates on a regular basis, and you should download and install those updates whenever they are available. When updates are available, a tiny little Windows icon appears in the *system tray* (the area in the lower-right corner of your screen, next to the clock). Click this icon to download and install updates. Alternatively, you can run Windows Update at any time by choosing Start⇨All Programs⇨Windows Update.

One more thing: If you buy a Pinnacle capture card, you may see a warning message during installation which says that the hardware installer hasn't passed testing for Windows XP compatibility. You'll see dire warnings about possible system instability, but I've found that it's safe to click Continue Anyway.

Using Hollywood FX Plug-ins for Pinnacle Studio

Like many more advanced (not to mention expensive) video-editing programs, Pinnacle designed Studio so it can be expanded and enhanced by using plug-ins. Pinnacle offers two different levels of the Hollywood FX plug-in:

- **Hollywood FX Plus:** This package includes 272 advanced transitions. You can also control lighting, trails, transparency, and many other aspects of objects on-screen. (I show you how to use some of these effects in Chapter 11.) Hollywood FX Plus retails for $49, and it comes free with Studio Deluxe.
- **Hollywood FX Pro:** This package provides an additional 96 transitions above and beyond what FX Plus offers, and includes other advanced special-effects tools. Hollywood FX Pro retails for $99.

Installing Hollywood FX is pretty simple. Just place the disc in your CD-ROM drive and follow the instructions on-screen to install. Once that's done, you can access Hollywood FX features in the regular Pinnacle Studio program window. Hollywood FX items will be enabled in Studio's transitions album. Figure D-1 shows the HFX Specialty Effects group in the transitions album. In the figure I have applied a Magic Ball transition to my project. Hollywood FX Plus is installed on this system, and the enabled effects are part of the Plus package. Effects with the word *Pro* across the front are not enabled because they only come with Hollywood FX Pro, which is not installed on this system.

Pinnacle offers other add-ons for Studio as well. In addition to the Hollywood FX plug-ins, you can also purchase additional DVD menus, video elements, and SmartSound music discs. To find Studio add-ons, click Help⇨Online Offers, and then click the Go Get It button in the dialog box that appears. Your Web browser should open to a list of product add-ons for Studio.

Figure D-1: Hollywood FX effects are enabled in the transitions album.

Appendix E

Using Windows Movie Maker

In This Appendix

- Upgrading Windows Movie Maker
- Getting familiar with Windows Movie Maker

Apple and Microsoft make the two most popular personal-computer operating systems in the world, so it should come as no surprise that they compete vigorously with each other. Early in the digital video boom, Apple released iMovie, introducing many Macintosh users to the world of moviemaking. Microsoft quickly followed suit with its own video-editing program, Windows Movie Maker. Windows Movie Maker was first released as part of the Windows Me (Millennium Edition) operating system, and was later included with Windows XP as well.

Like Apple iMovie, Windows Movie Maker is free for any Microsoft Windows user. But unlike iMovie, Windows Movie Maker doesn't get much coverage in this book. This appendix tells you why, and provides you with some basic information in case you do decide to use this program.

Microsoft buried Windows Movie Maker deep in the Windows Start menu, so it's not exactly the easiest thing to find. To launch Windows Movie Maker in XP, choose Start⇨All Programs⇨Accessories⇨Entertainment⇨Windows Movie Maker. In Windows Me, you'll find Windows Movie Maker in the Accessories menu.

Why Isn't Movie Maker Featured in This Book?

In theory, Windows Movie Maker is like many other free or low-cost video-editing programs. Movie Maker can capture video directly from your digital camcorder, and you can perform basic editing tasks — for example, assembling and cropping clips, adding transitions, making titles, and using a few other effects. But Windows Movie Maker has two important shortcomings when it comes to making movies:

- If you plan to record your movies on videotape or DVD, Windows Movie Maker's output quality is poor.
- Windows Movie Maker can only output movie files in Windows Media format, which requires Windows Media Player for playback. Although Windows Media Player is available free for Windows and Macintosh systems, many Mac users still don't have this program.

When you capture video from your digital camcorder using Windows Movie Maker, the software compresses your video to Windows Media format as it is captured. This means that right away the source footage that is stored on your hard disk is going to be of much lower quality. The plus side of course is that it takes up a lot less hard disk space. The highest capture quality that Windows Movie Maker allows uses less than 5 MB (megabytes) of disk space per minute. Compare that with conventional digital video capture in other programs like iMovie or Pinnacle Studio, which typically consumes about 200 MB per minute.

Of course, if you're only capturing video so you can put it online, the reduced quality is probably not a big deal. Movie Maker's only export format — Windows Media Video (WMV) — is great for online movies because it offers a pretty good balance of quality and file size. To view WMV-format movies, your audience must have Windows Media Player, which is available for both Windows and Macintosh systems. Still, if you wish you could export video in another format (such as RealVideo or MPEG), you won't be able to do it with Windows Movie Maker.

When it comes to making movies for my online friends, I actually really *like* Windows Movie Maker. In fact, I like it so much that I wrote *Microsoft Windows Movie Maker For Dummies* several years ago. Even though other programs — Pinnacle Studio (see Chapter 14) and Adobe Premiere, among others — can export in WMV format, none are as easy to use as Movie Maker. It's just that online movies are about all that Windows Movie Maker is good for; therefore, in this book, I don't feature it as much as Pinnacle Studio.

Upgrading Windows Movie Maker

The first version of Windows Movie Maker came with both Windows Me and Windows XP. Microsoft released version 2 of Movie Maker in early 2003, and if you haven't already upgraded to Windows Movie Maker 2, I strongly recommend it. Version 2 offers some welcome improvements, including

- **Real transitions:** Yes, it's true. In the original version of Movie Maker, your only available transition was a dissolve. But Windows Movie Maker 2 has 60 — yes, *sixty* — great transitions to choose from.
- **Video effects:** Windows Movie Maker offers 28 video effects that can be applied to your movies. That is 28 more effects than were available in Movie Maker version 1.
- **A title designer:** If you wanted to make titles for the first version of Windows Movie Maker, you had to do it using Windows Paint. Windows Movie Maker 2 includes a real title editor.

To upgrade your version of Windows Movie Maker, open the program and choose Help⇨Windows Movie Maker on the Web. This will take you to the Windows Movie Maker Web site, where you should see an Update link that takes you to a place where you can download the latest version. Alternatively, you can visit `www.microsoft.com/windowsxp/moviemaker/`.

Getting Acquainted with Windows Movie Maker

One thing that Windows Movie Maker definitely has going for it is a user interface that is really easy to understand and use. As does virtually every other video-editing program ever made, Movie Maker has a storyboard/timeline viewer at the bottom of the screen. The middle of the Movie Maker screen is filled with three items:

- **Movie tasks:** This panel is actually a collection of clickable menus covering all of the things you can do in Windows Movie Maker, including capturing video, using transitions and effects, and exporting the movie.
- **Browser:** The very middle of the screen serves as a clip browser if you're viewing a collection of clips, or it shows available transitions and effects if you've chosen to view those items by clicking them in the Movie Tasks panel.
- **Preview window:** No surprises here. The preview window contains playback controls that allow you to preview clips and your project.

As you can see in Figure E-1, Windows Movie Maker's timeline is actually quite similar to the timeline in Pinnacle Studio. The timeline includes tracks for video, audio, background music or narration, and titles. Movie Maker's timeline also provides a separate track for transitions. If you don't see the transition and audio track in your timeline, click the plus sign next to the video track.

Making movies with Windows Movie Maker is pretty straightforward. The Movie tasks panel provides a step-by-step process that you can follow to capture video, edit the movie, and finally export your finished project to a CD, the Internet, or a camcorder tape. When you're done editing a movie, click one of the export formats listed under Finish Movie and follow the simple on-screen instructions to export your project. For more information on using Windows Movie Maker, click the links under Movie Making Tips.

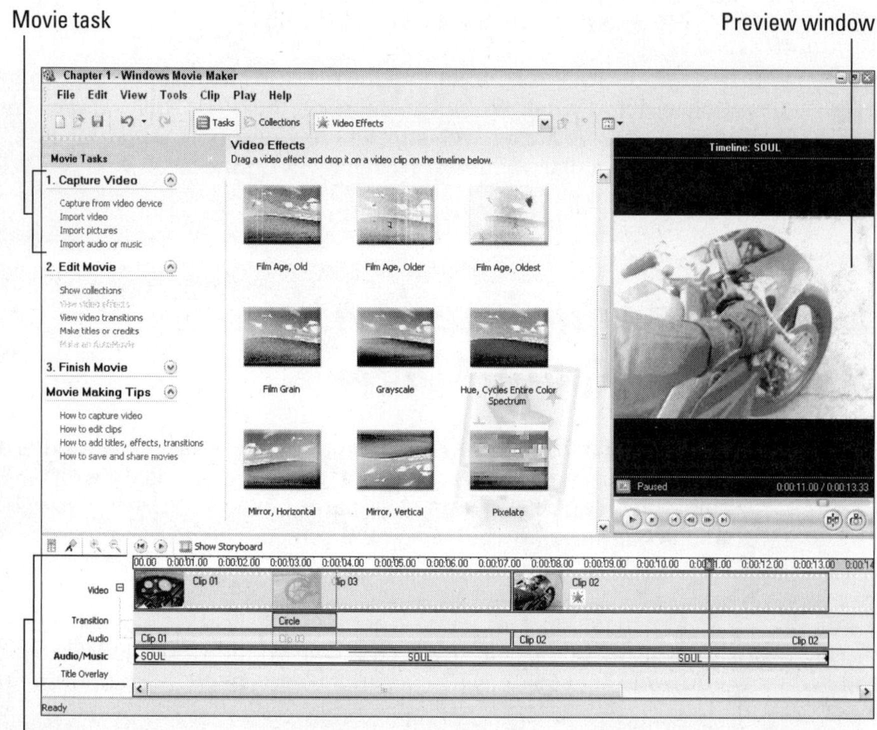

Figure E-1: Windows Movie Maker provides a friendly, easy-to-use interface.

Index

• Numbers •

• A •

• B •

• C •

• D •

• G •

• H •

• I •

• J •

• K •

• L •

• M •

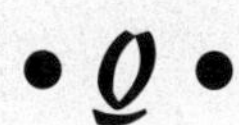

• T •

• U •

• V •

• W •

• X •

• Y •

• Z •

Notes

Notes

Notes

Notes

Notes

Wiley Publishing, Inc., End-User License Agreement

READ THIS. You should carefully read these terms and conditions before opening the software packet(s) included with this book "Book". This is a license agreement "Agreement" between you and Wiley Publishing, Inc. "WPI". By opening the accompanying software packet(s), you acknowledge that you have read and accept the following terms and conditions. If you do not agree and do not want to be bound by such terms and conditions, promptly return the Book and the unopened software packet(s) to the place you obtained them for a full refund.

1. **License Grant.** WPI grants to you (either an individual or entity) a nonexclusive license to use one copy of the enclosed software program(s) (collectively, the "Software," solely for your own personal or business purposes on a single computer (whether a standard computer or a workstation component of a multi-user network). The Software is in use on a computer when it is loaded into temporary memory (RAM) or installed into permanent memory (hard disk, CD-ROM, or other storage device). WPI reserves all rights not expressly granted herein.

2. **Ownership.** WPI is the owner of all right, title, and interest, including copyright, in and to the compilation of the Software recorded on the disk(s) or CD-ROM "Software Media". Copyright to the individual programs recorded on the Software Media is owned by the author or other authorized copyright owner of each program. Ownership of the Software and all proprietary rights relating thereto remain with WPI and its licensers.

3. **Restrictions on Use and Transfer.**

   (a) You may only (i) make one copy of the Software for backup or archival purposes, or (ii) transfer the Software to a single hard disk, provided that you keep the original for backup or archival purposes. You may not (i) rent or lease the Software, (ii) copy or reproduce the Software through a LAN or other network system or through any computer subscriber system or bulletin-board system, or (iii) modify, adapt, or create derivative works based on the Software.

   (b) You may not reverse engineer, decompile, or disassemble the Software. You may transfer the Software and user documentation on a permanent basis, provided that the transferee agrees to accept the terms and conditions of this Agreement and you retain no copies. If the Software is an update or has been updated, any transfer must include the most recent update and all prior versions.

4. **Restrictions on Use of Individual Programs.** You must follow the individual requirements and restrictions detailed for each individual program in the About the CD-ROM appendix of this Book. These limitations are also contained in the individual license agreements recorded on the Software Media. These limitations may include a requirement that after using the program for a specified period of time, the user must pay a registration fee or discontinue use. By opening the Software packet(s), you will be agreeing to abide by the licenses and restrictions for these individual programs that are detailed in the About the CD-ROM appendix and on the Software Media. None of the material on this Software Media or listed in this Book may ever be redistributed, in original or modified form, for commercial purposes.

5. **Limited Warranty.**

   (a) WPI warrants that the Software and Software Media are free from defects in materials and workmanship under normal use for a period of sixty (60) days from the date of purchase of this Book. If WPI receives notification within the warranty period of defects in materials or workmanship, WPI will replace the defective Software Media.

   (b) **WPI AND THE AUTHOR(S) OF THE BOOK DISCLAIM ALL OTHER WARRANTIES, EXPRESS OR IMPLIED, INCLUDING WITHOUT LIMITATION IMPLIED WARRANTIES OF MERCHANTABILITY AND FITNESS FOR A PARTICULAR PURPOSE, WITH RESPECT TO THE SOFTWARE, THE PROGRAMS, THE SOURCE CODE CONTAINED THEREIN, AND/OR THE TECHNIQUES DESCRIBED IN THIS BOOK. WPI DOES NOT WARRANT THAT THE FUNCTIONS CONTAINED IN THE SOFTWARE WILL MEET YOUR REQUIREMENTS OR THAT THE OPERATION OF THE SOFTWARE WILL BE ERROR FREE.**

   (c) This limited warranty gives you specific legal rights, and you may have other rights that vary from jurisdiction to jurisdiction.

6. **Remedies.**

   (a) WPI's entire liability and your exclusive remedy for defects in materials and workmanship shall be limited to replacement of the Software Media, which may be returned to WPI with a copy of your receipt at the following address: Software Media Fulfillment Department, Attn.: Digital Video For Dummies, 3rd Edition, Wiley Publishing, Inc., 10475 Crosspoint Blvd., Indianapolis, IN 46256, or call 1-800-762-2974. Please allow four to six weeks for delivery. This Limited Warranty is void if failure of the Software Media has resulted from accident, abuse, or misapplication. Any replacement Software Media will be warranted for the remainder of the original warranty period or thirty (30) days, whichever is longer.

   (b) In no event shall WPI or the author be liable for any damages whatsoever (including without limitation damages for loss of business profits, business interruption, loss of business information, or any other pecuniary loss) arising from the use of or inability to use the Book or the Software, even if WPI has been advised of the possibility of such damages.

   (c) Because some jurisdictions do not allow the exclusion or limitation of liability for consequential or incidental damages, the above limitation or exclusion may not apply to you.

7. **U.S. Government Restricted Rights.** Use, duplication, or disclosure of the Software for or on behalf of the United States of America, its agencies and/or instrumentalities "U.S. Government" is subject to restrictions as stated in paragraph (c)(1)(ii) of the Rights in Technical Data and Computer Software clause of DFARS 252.227-7013, or subparagraphs (c) (1) and (2) of the Commercial Computer Software - Restricted Rights clause at FAR 52.227-19, and in similar clauses in the NASA FAR supplement, as applicable.

8. **General.** This Agreement constitutes the entire understanding of the parties and revokes and supersedes all prior agreements, oral or written, between them and may not be modified or amended except in a writing signed by both parties hereto that specifically refers to this Agreement. This Agreement shall take precedence over any other documents that may be in conflict herewith. If any one or more provisions contained in this Agreement are held by any court or tribunal to be invalid, illegal, or otherwise unenforceable, each and every other provision shall remain in full force and effect.

FOR DUMMIES®

The easy way to get more done and have more fun

PERSONAL FINANCE

0-7645-5231-7

0-7645-2431-3

0-7645-5331-3

Also available:

Estate Planning For Dummies (0-7645-5501-4)
401(k)s For Dummies (0-7645-5468-9)
Frugal Living For Dummies (0-7645-5403-4)
Microsoft Money "X" For Dummies (0-7645-1689-2)
Mutual Funds For Dummies (0-7645-5329-1)
Personal Bankruptcy For Dummies (0-7645-5498-0)
Quicken "X" For Dummies (0-7645-1666-3)
Stock Investing For Dummies (0-7645-5411-5)
Taxes For Dummies 2003 (0-7645-5475-1)

BUSINESS & CAREERS

0-7645-5314-3

0-7645-5307-0

0-7645-5471-9

Also available:

Business Plans Kit For Dummies (0-7645-5365-8)
Consulting For Dummies (0-7645-5034-9)
Cool Careers For Dummies (0-7645-5345-3)
Human Resources Kit For Dummies (0-7645-5131-0)
Managing For Dummies (1-5688-4858-7)
QuickBooks All-in-One Desk Reference For Dummies (0-7645-1963-8)
Selling For Dummies (0-7645-5363-1)
Small Business Kit For Dummies (0-7645-5093-4)
Starting an eBay Business For Dummies (0-7645-1547-0)

HEALTH, SPORTS & FITNESS

0-7645-5167-1

0-7645-5146-9

0-7645-5154-X

Also available:

Controlling Cholesterol For Dummies (0-7645-5440-9)
Dieting For Dummies (0-7645-5126-4)
High Blood Pressure For Dummies (0-7645-5424-7)
Martial Arts For Dummies (0-7645-5358-5)
Menopause For Dummies (0-7645-5458-1)
Nutrition For Dummies (0-7645-5180-9)
Power Yoga For Dummies (0-7645-5342-9)
Thyroid For Dummies (0-7645-5385-2)
Weight Training For Dummies (0-7645-5168-X)
Yoga For Dummies (0-7645-5117-5)

Available wherever books are sold.
Go to www.dummies.com or call 1-877-762-2974 to order direct.

FOR DUMMIES®

A world of resources to help you grow

HOME, GARDEN & HOBBIES

0-7645-5295-3

0-7645-5130-2

0-7645-5106-X

Also available:

Auto Repair For Dummies (0-7645-5089-6)
Chess For Dummies (0-7645-5003-9)
Home Maintenance For Dummies (0-7645-5215-5)
Organizing For Dummies (0-7645-5300-3)
Piano For Dummies (0-7645-5105-1)
Poker For Dummies (0-7645-5232-5)
Quilting For Dummies (0-7645-5118-3)
Rock Guitar For Dummies (0-7645-5356-9)
Roses For Dummies (0-7645-5202-3)
Sewing For Dummies (0-7645-5137-X)

FOOD & WINE

0-7645-5250-3

0-7645-5390-9

0-7645-5114-0

Also available:

Bartending For Dummies (0-7645-5051-9)
Chinese Cooking For Dummies (0-7645-5247-3)
Christmas Cooking For Dummies (0-7645-5407-7)
Diabetes Cookbook For Dummies (0-7645-5230-9)
Grilling For Dummies (0-7645-5076-4)
Low-Fat Cooking For Dummies (0-7645-5035-7)
Slow Cookers For Dummies (0-7645-5240-6)

TRAVEL

0-7645-5453-0

0-7645-5438-7

0-7645-5448-4

Also available:

America's National Parks For Dummies (0-7645-6204-5)
Caribbean For Dummies (0-7645-5445-X)
Cruise Vacations For Dummies 2003 (0-7645-5459-X)
Europe For Dummies (0-7645-5456-5)
Ireland For Dummies (0-7645-6199-5)
France For Dummies (0-7645-6292-4)
London For Dummies (0-7645-5416-6)
Mexico's Beach Resorts For Dummies (0-7645-6262-2)
Paris For Dummies (0-7645-5494-8)
RV Vacations For Dummies (0-7645-5443-3)
Walt Disney World & Orlando For Dummies (0-7645-5444-1)

Available wherever books are sold. Go to www.dummies.com or call 1-877-762-2974 to order direct.

FOR DUMMIES®

Plain-English solutions for everyday challenges

COMPUTER BASICS

0-7645-0838-5

0-7645-1663-9

0-7645-1548-9

Also available:

PCs All-in-One Desk Reference For Dummies (0-7645-0791-5)

Pocket PC For Dummies (0-7645-1640-X)

Treo and Visor For Dummies (0-7645-1673-6)

Troubleshooting Your PC For Dummies (0-7645-1669-8)

Upgrading & Fixing PCs For Dummies (0-7645-1665-5)

Windows XP For Dummies (0-7645-0893-8)

Windows XP For Dummies Quick Reference (0-7645-0897-0)

BUSINESS SOFTWARE

0-7645-0822-9

0-7645-0839-3

0-7645-0819-9

Also available:

Excel Data Analysis For Dummies (0-7645-1661-2)

Excel 2002 All-in-One Desk Reference For Dummies (0-7645-1794-5)

Excel 2002 For Dummies Quick Reference (0-7645-0829-6)

GoldMine "X" For Dummies (0-7645-0845-8)

Microsoft CRM For Dummies (0-7645-1698-1)

Microsoft Project 2002 For Dummies (0-7645-1628-0)

Office XP For Dummies (0-7645-0830-X)

Outlook 2002 For Dummies (0-7645-0828-8)

Get smart! Visit www.dummies.com

- **Find listings of even more *For Dummies* titles**
- **Browse online articles**
- **Sign up for Dummies eTips™**
- **Check out *For Dummies* fitness videos and other products**
- **Order from our online bookstore**

Available wherever books are sold. Go to www.dummies.com or call 1-877-762-2974 to order direct.

FOR DUMMIES®

Helping you expand your horizons and realize your potential

INTERNET

0-7645-0894-6

0-7645-1659-0

0-7645-1642-6

Also available:

America Online 7.0 For Dummies (0-7645-1624-8)

Genealogy Online For Dummies (0-7645-0807-5)

The Internet All-in-One Desk Reference For Dummies (0-7645-1659-0)

Internet Explorer 6 For Dummies (0-7645-1344-3)

The Internet For Dummies Quick Reference (0-7645-1645-0)

Internet Privacy For Dummies (0-7645-0846-6)

Researching Online For Dummies (0-7645-0546-7)

Starting an Online Business For Dummies (0-7645-1655-8)

DIGITAL MEDIA

0-7645-1664-7

0-7645-1675-2

0-7645-0806-7

Also available:

CD and DVD Recording For Dummies (0-7645-1627-2)

Digital Photography All-in-One Desk Reference For Dummies (0-7645-1800-3)

Digital Photography For Dummies Quick Reference (0-7645-0750-8)

Home Recording for Musicians For Dummies (0-7645-1634-5)

MP3 For Dummies (0-7645-0858-X)

Paint Shop Pro "X" For Dummies (0-7645-2440-2)

Photo Retouching & Restoration For Dummies (0-7645-1662-0)

Scanners For Dummies (0-7645-0783-4)

GRAPHICS

0-7645-0817-2

0-7645-1651-5

0-7645-0895-4

Also available:

Adobe Acrobat 5 PDF For Dummies (0-7645-1652-3)

Fireworks 4 For Dummies (0-7645-0804-0)

Illustrator 10 For Dummies (0-7645-3636-2)

QuarkXPress 5 For Dummies (0-7645-0643-9)

Visio 2000 For Dummies (0-7645-0635-8)

Available wherever books are sold. Go to www.dummies.com or call 1-877-762-2974 to order direct.

FOR DUMMIES®

The advice and explanations you need to succeed

SELF-HELP, SPIRITUALITY & RELIGION

0-7645-5302-X

0-7645-5418-2

0-7645-5264-3

Also available:

- The Bible For Dummies (0-7645-5296-1)
- Buddhism For Dummies (0-7645-5359-3)
- Christian Prayer For Dummies (0-7645-5500-6)
- Dating For Dummies (0-7645-5072-1)
- Judaism For Dummies (0-7645-5299-6)
- Potty Training For Dummies (0-7645-5417-4)
- Pregnancy For Dummies (0-7645-5074-8)
- Rekindling Romance For Dummies (0-7645-5303-8)
- Spirituality For Dummies (0-7645-5298-8)
- Weddings For Dummies (0-7645-5055-1)

PETS

0-7645-5255-4

0-7645-5286-4

0-7645-5275-9

Also available:

- Labrador Retrievers For Dummies (0-7645-5281-3)
- Aquariums For Dummies (0-7645-5156-6)
- Birds For Dummies (0-7645-5139-6)
- Dogs For Dummies (0-7645-5274-0)
- Ferrets For Dummies (0-7645-5259-7)
- German Shepherds For Dummies (0-7645-5280-5)
- Golden Retrievers For Dummies (0-7645-5267-8)
- Horses For Dummies (0-7645-5138-8)
- Jack Russell Terriers For Dummies (0-7645-5268-6)
- Puppies Raising & Training Diary For Dummies (0-7645-0876-8)

EDUCATION & TEST PREPARATION

0-7645-5194-9

0-7645-5325-9

0-7645-5210-4

Also available:

- Chemistry For Dummies (0-7645-5430-1)
- English Grammar For Dummies (0-7645-5322-4)
- French For Dummies (0-7645-5193-0)
- The GMAT For Dummies (0-7645-5251-1)
- Inglés Para Dummies (0-7645-5427-1)
- Italian For Dummies (0-7645-5196-5)
- Research Papers For Dummies (0-7645-5426-3)
- The SAT I For Dummies (0-7645-5472-7)
- U.S. History For Dummies (0-7645-5249-X)
- World History For Dummies (0-7645-5242-2)

Available wherever books are sold. Go to www.dummies.com or call 1-877-762-2974 to order direct.

FOR DUMMIES®

We take the mystery out of complicated subjects

WEB DEVELOPMENT

0-7645-1643-4

0-7645-0723-0

0-7645-1630-2

Also available:

ASP.NET For Dummies (0-7645-0866-0)
Building a Web Site For Dummies (0-7645-0720-6)
ColdFusion "MX" For Dummies (0-7645-1672-8)
Creating Web Pages All-in-One Desk Reference For Dummies (0-7645-1542-X)
FrontPage 2002 For Dummies (0-7645-0821-0)
HTML 4 For Dummies Quick Reference (0-7645-0721-4)
Macromedia Studio "MX" All-in-One Desk Reference For Dummies (0-7645-1799-6)
Web Design For Dummies (0-7645-0823-7)

PROGRAMMING & DATABASES

0-7645-0746-X

0-7645-1657-4

0-7645-0818-0

Also available:

Beginning Programming For Dummies (0-7645-0835-0)
Crystal Reports "X" For Dummies (0-7645-1641-8)
Java & XML For Dummies (0-7645-1658-2)
Java 2 For Dummies (0-7645-0765-6)
JavaScript For Dummies (0-7645-0633-1)
Oracle9*i* For Dummies (0-7645-0880-6)
Perl For Dummies (0-7645-0776-1)
PHP and MySQL For Dummies (0-7645-1650-7)
SQL For Dummies (0-7645-0737-0)
VisualBasic .NET For Dummies (0-7645-0867-9)
Visual Studio .NET All-in-One Desk Reference For Dummies (0-7645-1626-4)

LINUX, NETWORKING & CERTIFICATION

0-7645-1545-4

0-7645-0772-9

0-7645-0812-1

Also available:

CCNP All-in-One Certification For Dummies (0-7645-1648-5)
Cisco Networking For Dummies (0-7645-1668-X)
CISSP For Dummies (0-7645-1670-1)
CIW Foundations For Dummies with CD-ROM (0-7645-1635-3)
Firewalls For Dummies (0-7645-0884-9)
Home Networking For Dummies (0-7645-0857-1)
Red Hat Linux All-in-One Desk Reference For Dummies (0-7645-2442-9)
TCP/IP For Dummies (0-7645-1760-0)
UNIX For Dummies (0-7645-0419-3)

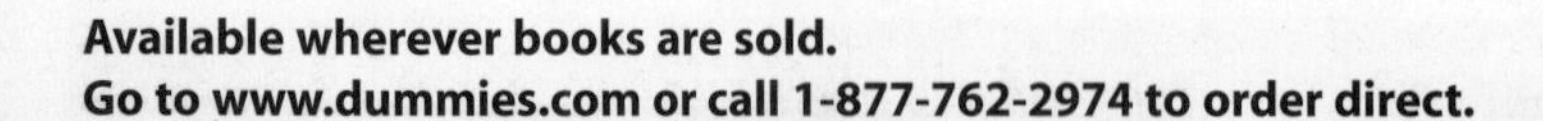